PROXIES

INFRASTRUCTURES SERIES

edited by Geoffrey C. Bowker and Paul N. Edwards

A list of books in the series appears at the back of the book.

PROXIES

The Cultural Work of Standing In

DYLAN MULVIN

The MIT Press
Cambridge, Massachusetts
London, England

The open-access edition of this book was made possible by generous funding from Arcadia—a charitable fund of Lisbet Rausing and Peter Baldwin.

This book was set in Adobe Garamond Pro and Berthold Akzidenz Grotesk by Westchester Publishing Services. Printed and bound in the United States of America.

Library of Congress Cataloging-in-Publication Data

Names: Mulvin, Dylan, author.
Title: Proxies : the cultural work of standing in / Dylan Mulvin.
Description: Cambridge, Massachusetts : The MIT Press, [2021] |
    Series: Infrastructures series | Includes bibliographical references and index.
Identifiers: LCCN 2020050161 | ISBN 9780262045148 (paperback)
Subjects: LCSH: Simulation methods—History. | Standardization—Social aspects—
    History. | Analogy.
Classification: LCC T57.62 .M85 2021 | DDC 003—dc23
LC record available at https://lccn.loc.gov/2020050161

10 9 8 7 6 5 4 3 2 1

for Alice
for Ida

Lauren Hoffmann, Tom LaMarre, Cait McKinney, Sharif Mowlabocus, Andrew Piper, and Fred Turner. Thank you, as well, to Robin Lynch for the research correspondence.

At Microsoft Research, Nancy Baym, Tarleton Gillespie, and Mary Gray have created their own unique place for inquiry and intellectual cross-pollination. For two years, I had the ridiculous privilege of thinking through this project, and many other things, with them, Dan Greene, and our residents and visitors. There, and elsewhere, I am grateful for conversations with Meryl Alper, Mike Ananny, Dalida María Benfield, Rena Bivens, Jack Bratich, André Brock, Robyn Caplan, Nick Couldry, Kate Crawford, Stephanie Dick, Kevin Driscoll, Stefanie Duguay, Michaelanne Dye, Nathan Ensmenger, Megan Finn, Ysabel Gerrard, Jack Gieseking, Sarah T. Hamid, Caroline Jack, Steve Jackson, Elena Maris, Shannon McGregor, Mara Mills, Sue Murray, Lisa Parks, Genevieve Patterson, Chris Persaud, Jill Walker Rettberg, Craig Robertson, Nick Seaver, Hannah Spaulding, Lana Swartz, T. L. Taylor, Bill Thies, Nanna Bonde Thylstrup, Penny Trieu, Anjalie Vats, Kelly Wagman, Meredith Whitaker, Ben Woo, and Ming Yin.

The students at the London School of Economics (LSE) have inspired many new ideas, and I can't wait to work on the next project with you. And I am fortunate to be surrounded by inspiring and warm colleagues in London. I am grateful to Sarah Banet-Weiser, for making my landing easier and helping me find the space to finish this work. My colleagues set an exceptionally high bar for their work, but, more important, they have provided friendship at a time when it was sorely needed. My gratitude goes especially to Nick Anstead, Omar Al-Ghazzi, Cath Bennett, Bart Cammaerts, Julia Corwin, James Deeley, Simidele Dosekun, Lee Edwards, Seeta Peña Gangadharan, Nicole Garnier, Ellen Helsper, Sonia Livingstone, Bingchun Meng, Jean-Christophe Plantin, and Alison Powell for helping me make London home.

Audiences at the University of Toronto, University of Indiana, University of Colorado, Denver, Harvard, Northwestern, and the Copenhagen Business School asked hard questions and helped me refine this project over the years. Thank you to the University of Maryland and Matthew Kirschenbaum for hosting me during my research in DC.

Acknowledgments

My work is indebted to the care and curiosity of my friends, family, teachers, and colleagues. This work began at McGill University, where Jonathan Sterne and Carrie Rentschler have created a home for humane and original scholarship. Jonathan is my mentor, collaborator, and friend and he demonstrates how kindness is possible within academia and beyond. I strive to be for others what he is to so many people in our community. Carrie is the best teacher and reader I have known, and I am a better thinker and person for knowing her, her power, her compassion, and her lack of bullshit. Carrie showed me that this was a project about labor.

The Department of Art History and Communication Studies is a unique place to study. Thank you to Jenny Burman, Will Straw, and Darin Barney for their teaching and service to that academic community, and to Darin for never once letting me win at squash. At SFU, Zoë Druick provided guidance and mentorship when it was needed, and I am always in her debt.

At McGill, I was surrounded by amazing people and am lucky to count Joceline Andersen, Anna Candido, Li Cornfeld, Paul Fontaine, Amy Hasinoff, Jess Holmes, Rafico Ruiz, and Abi Shapiro among my friends. From Somerville–Cambridge, Devin Kennedy, Gili Vidan, and Marc Aidinoff have become crucial correspondents about, basically, everything.

A special appreciation is owed to the people who read parts of this work charitably and discovered facets that were hidden to its author. Thank you to Nancy Baym, Biella Coleman, T. L. Cowan, Tarleton Gillespie, Anna

Contents

There is no scholarly work without the people who work to maintain the traces of history in archives, libraries, and institutions. I need to thank the staff who let me in their doors and were always patient with my admittedly strange requests. Equally, I am thankful for the support staffs at McGill, Microsoft Research, and the LSE for their knowledge, expertise, and consideration. This work was also aided by the financial support of the Social Sciences and Humanities Research Council of Canada, Media@McGill, and the Wolfe Chair in Scientific and Technological Literacy. This project was also completed because of the life-altering labor of the people at Bigelow Cooperative Daycare, Osprey Daycare, LSE Nursery, and Heathbrook Primary, as well as Yulia Salnikova, Trista Charron, and Taylor Doyle.

At the MIT Press, I have had the pleasure of working with Katie Helke and Justin Kehoe, who brought this project into the press and urged it over the finish line. My gratitude goes to them, to the three very generous and persuasive anonymous reviewers, and to Geof Bowker and Paul Edwards. Final manuscript assistance came from Meichen Waxer, indexing from Jason Begy, and photography and cardboard skills from Lewis Bush.

My parents, Tania and Robert, made it possible for me to be a student for several decades—I am grateful for their love and care. Thank you to Jesse and Stephanie, and the cousins, for your love from afar; and to Susan Christensen for all of your love and support. The friendship of Lee, Karl, and Dominique makes me homesick—and of Tali and Brian, homesick for that other place. Thank you to Tarleton for all the rides and everything else. Thank you, Sharif, for friendship and much more. Thank you R. R., for all the art, both in this book and everywhere else.

Cait McKinney is a great friend, teammate, and intellect and is somehow one of my oldest friends and my closest collaborator. Every bit of this project was refracted through our conversations. Thanks, C.

Alice and Ida, thank you for making these lives together.

1 SAMPLES OF THE WORLD *OUT THERE*: THE SURROGATE LOGIC OF PROXIES

What, then, is contained in the *as if*?

—HANS VAIHINGER[1]

YODAVILLE

There is a small town in the desert of Arizona's southwestern edge, thirty-five miles southeast of Yuma and fewer than six miles from the US-Mexico border. The buildings in the town, mostly one story, are distributed throughout small neighborhoods and laid out across eight axes. There is one road in and out of town. The town's official name is Urban Target Complex (R-2301-West), but everyone knows the place by its nickname, "Yodaville," a name given to honor Floyd "Yoda" Usry, a now-retired colonel from the US Marine Corps. No one lives in Yodaville because Yodaville was never meant to be lived in. Instead, it is a fabricated test city, a bombing range constructed in the late 1990s to train Marine Corps pilots in attacking cities and supporting ground troops at war in urban settings. The houses and buildings of this test city are made of shipping containers and empty bomb canisters. The inhabitants are stick figures made of metal.[2]

In the early 1990s, during the intervention of the United Nations (UN) in the Somali Civil War, the US military undertook an operation in the capital city of Mogadishu. The operation ended with hundreds of Somali casualties and nineteen deaths among American special forces. Memorialized in 1999 by journalist Mark Bowden's *Black Hawk Down: A*

Story of Modern War and a 2001 Hollywood film based on that book, the battle of Mogadishu is widely regarded as a failure of the US military to prepare for a new kind of urban warfare.[3] Yodaville was built by the military to respond to the failures of Mogadishu.

In a RAND report on urban warfare from 2006, the authors begin with the American deaths in Mogadishu.[4] They draw a direct connection between the battle and the construction of Yodaville:

> The desperate October 1993 fighting on the streets of Mogadishu triggered U.S. Army development of a new type of urban training facility, *one designed to be less like the pristine villages of northwest Europe and more akin to the chaotic environments found in densely populated areas of the developing world.* The Marine Corps built "Yodaville," an innovative training site in Arizona that vividly replicates the difficulties of engaging urban targets from aircraft.[5]

The American deaths in Mogadishu (the Somali deaths go unmentioned) were partly attributable to their military training, and as the RAND report argues, this was a symptom of institutional failures of imagination in the design of training simulations. The implication is that these were not poor fighters; instead, they were people trained on untimely representations of faraway threats designed for a bygone era. The soldiers could not picture their new battleground because their references were askew. Having based their simulated fighting on the "pristine" architecture of northwest Europe, the US military was unprepared for the landscapes (to use RAND's terminology) of a "chaotic" and developing world. The remedy came in the form of a new training site that was meant to be more akin to the sites of conflict in the changing landscape of the American empire.[6]

By adjusting the environment of their training site to a not-yet-named-but-immanently-chaotic elsewhere, the military hoped that their new simulations would be commensurate with the likely arenas of future conflict and the embodied experience of targeting, attacking, and escaping those places. The look, shape, and feel of European villages were inscribed in the institutional memory of the military's training protocols, so a disaster like the battle of Mogadishu forced a rupture with the past; what followed was a new set of inscriptions.[7] The battle of Mogadishu left traces that were archived in the building of Yodaville (figure 1.1), which acts as both a memorial to past failures and a fortification against future ones.

In the attempt to recalibrate US imperialism after a moment of breakdown, the US Marine Corps built a fake town to replicate the imagined shape, texture, and feel of an emergent enemy territory. Once built, Yodaville quickly became a *proxy* city for simulating a new kind of warfare. It was a territory for practicing the enforcement of an empire and the changing character of that empire's boundaries. As a proxy, Yodaville could represent and materialize this new enemy territory through a basic logic: by approximating emergent combat zones, the armed forces could act as if they were in combat. Soldiers could be trained and tested on surrogate targets that reflected their eventual ones, and a new standard of combat could be established, learned, and embodied.

Figure 1.1
An artist's rendering of a frame from the video *Marines Fire on Yodaville* (2015), viewed on Military.com. Yodaville is seen here from behind the shoulder of a gunner aboard a US Marine Corps UH-1Y Venom gunship. The buildings (shipping containers) radiate out from the center of Yodaville. Image: R. R. Mulvin.

PROXIES

Yodaville is a proxy for any number of places in the world. This simulated town was constructed to be a flexible stand-in for many potential places (emergent battlegrounds) while still being coherently a single place (Yodaville) so that it could be used as a shared testing site for training and evaluation. Placed pointedly on the border of the continental United States, Yodaville is the materialized stand-in for a limited and controllable enemy territory. It is part of a chain of US military encampments along the US-Mexico border, it is situated within the occupied traditional territories of the Quechan and Cocopah, and it adjoins the Fort Yuma Indian Reservation.[8] As such, Yodaville is used as a stand-in while simultaneously serving as an instrument for sustained foreign and domestic occupation. By serving this double function, it embodies a very specific fantasy: "to become without becoming,"[9] meaning it can, momentarily, become a foreign war zone without risking any of its actual territorial occupations. This is the promise of proxies and the promise of creating controllable renditions of an unpredictable and unknowable world.

Proxies function as the necessary forms of make-believe and surrogacy that enable the production of knowledge. Such knowledge production relies on accessible representations of the world, and proxies are the people, artifacts, places, and moments invested with the authority to represent the world. To interrogate the use of proxies is to ask: *to whom or to what do we delegate the power to represent the world?* To answer that question, I trace the lives of long-lasting, entrenched, and thoroughly standardized delegates—proxies—and the cultural work that people undertake to act as stand-ins and keep these stand-ins viable. This includes the moments of genesis, when communities of practitioners ask themselves what they would use as a delegate for the world, and the moments, later in the lives of these proxies, when the use of alternative delegates seems unimaginable. This is the pathway by which a proxy becomes common sense. People work with proxies to produce knowledge, maintain measurement systems, evaluate performance, and engage in a series of practices that are made possible only by investing certain materials with the power to represent an unpredictable world.[10]

As a component of the internetwork of knowledge infrastructures, *proxification* is a culturally conditioned practice of consistently using some things to stand in for the world.[11] Proxies are intermediaries—they mediate between the practicality of getting work done and the collective, aesthetic, and political work of capturing the world in an instant. As Yodaville attests, the choice and development of a proxy for the world often constitute an attempt to wrangle and control the unpredictable. Every proxy comes to exist in singular ways and represents one method that experts have used to evoke a world and, by evoking a world, bring it to be. But even where proxies have idiosyncratic origins, they share in their reliance on the cultural work of standing in. Culture consists, following Marilyn Strathern, of the ways that people draw analogies between things, "in the way certain thoughts are used to think others."[12] This definition ought to orient us toward proxies as analogies, as the material for making connections (as the stuff for thinking), and the ways in which such materials both animate communities and reveal their exclusions. The cultural work of standing in, then, is the work of both analogizing and maintaining the relationship between that which *is* and that which works *as if*.

Almost like religious relics, proxies are saturated in meaning—and their further use only reinvigorates the idea that these things, these people, and these places are special: they are imbued with the power to stand in. Through the stories of three proxies that have historically evoked this enchantment—the International Prototype Kilogram (IPK), the "Lena" test image, and the standardized patient program—this book traces the ways that communities of scientific and technical professionals engage in the theatrical enactment of objectivity through the embodied use and upkeep of proxies. We will look at the guidelines for how much manual pressure to employ when scrubbing an official kilogram clean; the frantic moment when an electrical engineer tore a centerfold from a *Playboy* magazine to create a now-canonical test image for digital image processing; and the ways that medical actors (so-called standardized patients) are trained to embody the typical symptoms of diseases in order to train physicians.

By beginning with strict bodily protocols for cleaning kilograms and ending with protocols for transforming human bodies to make them better stand-ins, this book charts a path from the eighteenth century to the near-present. It begins with a moment in which a belief took hold that the natural sciences could expose so-called invariants of nature and use them as the basis of universal standards; it ends with the twentieth-century belief that illness could be adequately codified such that it could be reproduced in a performer. These proxies are vital to the work of standardizing knowledge, and they themselves also become standardized, eventually entrenching as infrastructural and pregiven conditions for making sense of the world. But this work never happens in a vacuum: proxies shape and are shaped by the politics of representation and delegation. Test images have historically reproduced a racist and sexist visual culture that codes white femininity as a prototype; standardized patients rely on actors who wear disability as a masquerade; and these standards, in turn, shape the capacities that people have to build their own worlds.

Standardization is a process of forgetting. As Andrew Russell says, standardization is "a social process by which humans come to take things for granted."[13] Just as we could not imagine our world without its fundamental systems (including the metric system, digital image transmission, and the medical profession), it is impossible to imagine these systems without the use of their proxies. Where people share common references, to exchange knowledge and compare experiences of the world, they will produce and maintain proxies. But if the success of standardization is marked by forgetting the work of documenting the process requires one to recover the memory of how we got here. This examination is built on the idea that any notion of the "circuit of culture" ought to include practices of standardization, in addition to the conventional nodes of identity, production, representation, regulation, and consumption.[14] Standardization shows how ideas are formalized, but it also takes place in a cultural context in which those involved are themselves consumers whose identity positions bear on their work.

> > >

What is required to take a proxy for granted? Think of Yodaville: for it to stand in for the many potential battlefields of the US military, one must take for granted that the United States occupies the territory known as Arizona; that it can do with it what it pleases (including using it as a staging ground of simulated warfare); that the deserts of Arizona are similar enough to the deserts of other places to provide sufficient comparison; and that one soldier's performance in that desert is comparable to another's. These are the suspensions of disbelief that can transform some shipping containers outside Yuma into a durable stand-in for a world of potential battlegrounds.

We can recognize proxies all around us. When we invest something with the power to represent something else, we are engaged in the logics of surrogacy and vicarity, however minor. Cut out a template for a sewing project and you've made a ready-to-hand proxy; adjust your television or computer monitor using color bars and you've used a built-in proxy for the broadly expected formal qualities of digital screens. We have other, familiar proxies too: the proxy vote of a shareholder, a power of attorney document, and the proxy logic of representative democracy; these are all standardized and complex apparatuses for designating other human beings as our proxies in key moments of decision-making. These people, documents, and things are trusted delegates that social convention allows to serve as surrogates.

Delegation is a vital part of how institutions survive and constitutes one of the most basic relationships between people and technology.[15] For Bruno Latour, for instance, people delegate labor to nonhumans to script desired results: concrete speed bumps are delegated a function of police officers when they are enlisted with the goal of slowing down traffic.[16] But a speed bump is never only a speed bump, and neither is it only a lump of concrete made to condition behavior.[17] Delegation is a primary means of displacing social relations, and the delegate is, inescapably, an artifact of those relations.[18] Only some people possess the power to delegate core functions of the state to things, and marginalized and minoritized populations are more likely to be the targets of control: whether it's the vision of the state delegated to a security camera or a person's prospects for employment delegated to a credit score. Delegation is a political means of distributing the possibilities of living a flourishing and secure life.[19]

Delegation is a fragile process, entirely dependent on a network of trust between the delegator and the delegate (will you be a faithful representative?), as well as those with an interest in the power of a proxy to stand in for its counterpart. Through the histories of proxies, I show how people, places, and things come to be taken for granted, what happened when they were challenged, and how their trustworthiness as delegates was once again rebuilt. By building a catalog of proxies that centers on human bodies and human labor, I make the following arguments:

- *Proxies are bodily*: this is visible in the work of measurement and training that relies on finely tuned embodied and relational practices.
- *Proxies are both sticky and permeable*: though proxies are built as labor-saving devices to stand in for worldly phenomena, they inevitably carry and leave traces of their cultural milieus and the places where they've traveled.
- *Proxies rely on suspended disbelief*: the scientific and technical expertise underlying them is formed and repeated through scenes of performance, where participants must act as if a stand-in were the real thing for the purposes of getting work done.

THE MATERIAL LIFE OF STAND-INS

"All sciences must deal with the problem of selecting and constituting 'working objects,'" write Lorraine Daston and Peter Galison.[20] And proxies are no different. As simplified representations meant to be pregiven, ready at hand, easily called into action, and unquestioned, proxies are subject to many of the rules of working objects. The history of science and technology is strewn with studies of working objects, from the use of the fruit fly *Drosophila melanogaster* in genetics research, to the use of mice in biomedical research, to the use of chess in artificial intelligence testing, to the use of human surrogates in automobile crash testing.[21] All these examples are meant, ultimately, as practical solutions to the problem of needing usable models of a "too plentiful and too various" world.[22] As Steven Shapin writes, "All testimony about states of affairs stands in a metonymic relationship to those states of affairs, and the condition of your knowing about these things—otherwise unavailable to you—is your accepting the legitimacy of that relationship."[23] A geneticist probably doesn't have any natural affinity for fruit flies but may have an affinity for the community they share with people who work with fruit flies, as well as an investment in the idea that fruit flies can stand in for (some) other living organisms: "the local and the specific are not the *point* of these experiments . . . but in order for specific findings to be *about* the atmosphere or *about* the universe the credibility of these standing-for relationships have to be accepted."[24] In other words, the legitimacy of an experiment, a model organism, or a statement about how the world works hinges on both a metonymic relationship (i.e., something stands in for something else) and the credibility of the stand-ins (i.e., someone believes it) to act for other phenomena that are not or could not be made present.

And yet: the local and the specific *do* matter, and no part of this arrangement is simple. The bonds of cultural, social, and professional norms are pulled taut every time an equivalence is made. We can examine these bonds over time and between spaces to investigate the tension between the thing standing in and the thing being stood for, or the person making the connection and the person being asked to believe it. Proxies are the product of creative decisions to design and maintain trusted delegates of a too-plentiful and various world, and a focus on the cultural labor of standing in shifts the analytic emphasis from the singular choice of a potential metonym (e.g., a fly is like other living organisms) to the training and the ongoing work that maintain proxies as credible and, ultimately, indispensable.

Proxies are instrumental to developing "group-licensed ways of seeing,"[25] and they are crucial to the ways we learn how to participate in our communities by training ourselves through common references, by coming to see problems as akin, and by taking for granted that others in our community share those references and those ways of seeing. Broadly, this process has many artifacts—tacit knowledge, canons, hidden curricula, inside jokes[26]—only some of which take the material form of common instruments for knowledge production. When Thomas Kuhn revised *The Structure of Scientific Revolutions*, he added a new focus on "exemplars" to show how membership in a scientific community could not be explained merely through a shared set of rules. He used these exemplars to respecify what he meant by a scientific paradigm: "Shared examples of successful practice could . . . provide what the group lacked in rules. Those examples were its paradigms, and as such essential to its continued research."[27]

In a related fashion, Michelle Murphy calls particular representations of quantification like graphs of gross domestic product (GDP) *phantasmagrams*, speaking to the power of such instruments to far surpass their mere utility. The phantasmagraphic power of some quantitative practices means that they are "enriched with affect, propagate imaginaries, lure feeling, and hence have supernatural effects in surplus of their rational precepts."[28] Exemplars, working objects, and phantasmagrams, though each is distinct, are kindred ways of understanding how surrogate logics shape and bind disciplinary communities: they work through articulation, by drawing connections between things, through routine and practice, and through the shared bonds that communities form with particular problems and examples. But these concepts risk placing too much emphasis on the "objects" of laboratories and classrooms and not enough emphasis on the labor and affective commitments that proxies inspire. They also risk displacing the ways that human bodies must carry the traces of this work, either through the repetitive use of a narrow set of exemplars or the psychological and physical toll that the work might take. For instance, test images (discussed in chapters 3 and 4) are often singled out for their representational injustices on gendered and racialized grounds. But as compulsory instruments used in scientific, industrial, and classroom settings, their users often have little power to refuse their use or to question their credibility as stand-ins.

The "objectness" of working objects also appears especially brittle when we consider how many shared proxies are alive. Not just flies, rats, and mice, but living humans who work as test subjects, model patients, or make up case studies. Take the standardized patient program (discussed in chapter 5), in which laypersons are taught to embody the normal symptoms of diseases that they don't have in order to train and test physicians in diagnostic techniques. As proxies, these individuals must suppress their own idiosyncrasies (including any actual maladies they might carry) to elevate their common status as humans with the capacity for sickness. Not only would it be dehumanizing to describe standardized patients as "working objects," it would fail to account for the ways that choosing and maintaining shared proxies entail much more than agreeing on an adequately typical exemplar. It would miss the sinuous work that analogies perform and it would erase the human labor involved in working as a stand-in. Or take a more conventional example from the world of science: chapter 2 deals at length with the protocols for cleaning and washing the IPK by hand. For more than a century, the IPK stood in for the concept of mass across the globe. In many ways, the IPK and other standard kilograms are classic working objects: platinum-iridium facsimiles of a cubic decimeter of water, they are made to be both inordinately precise and readily accessible. Yet, to be credible stand-ins, they require highly choreographed, manual cleaning with an ether-ethanol solution and a chamois cloth. This cleaning is so fundamental to the viability of the IPK as a standard that it was added to the official definition of mass within the metric system.

What distinguishes the histories of proxies in this book from histories of working objects, then, is the focus on embodied labor and performance as indispensable to the maintenance of knowledge infrastructures.[29] The cleaning of the IPK was not supplemental to the meaning of mass within the metric system; it is fundamental to what made it a viable standard. A larger set of affective and cultural practices binds people to their proxies as compulsory tools, binds communities to their shared representations, and tasks other people with the labor of making, using, and maintaining those representations.

As proxies travel to new sites and persist as "interscalar vehicles," the arbitrariness of their relationship to the world *out there* can appear in stark relief.[30] Seen in the wrong light by an ungenerous audience, what once seemed like a credible stand-in for the world starts to look threadbare, anachronistic, idiosyncratic, or outright unjust. There is no natural correspondence between shipping containers in southwestern Arizona and any potential target of the US military. It is people who must constantly reassert that correspondence, agree to it, and keep it coherent. It is also people who must embody that correspondence, either as the workers tasked with making and maintaining standards and infrastructures or as people who must make do within conditions not of their own choosing.[31] I therefore follow an approach to infrastructural labor that is attuned to the ways that infrastructure is manifested in human practices. The knowledge labor that I trace is material and affective work, constituted in practices that, as Cait McKinney puts it, "assemble people, information, and technologies toward social goals."[32] And to follow Jacqueline Wernimont, this is an attempt to rematerialize test data, "to make it into something that one can touch, feel, own, give, share, and spend time with."[33] This is all to say that my approach is interested in the materialization of ideas in things and assumes that things are ineluctably made up of relationships.

A WORLD OF PROXIES

I live in England, but there are times that I might need to watch Canadian television. A virtual private network (VPN) can make my computer appear as if it was in Canada when contacting Canadian servers—it does this by masking my Internet Protocol (IP) address through something called a "proxy server." Because IP addresses are often a trusted stand-in for location, I can exploit a network of makeshift signifiers to bypass geofenced content (hypothetically). When Ari Luotonen and Kevin Altis published their foundational 1994 paper on web proxies, they introduced the technology through the labor-saving potential of proxies: "A proxy *allows client writers to forget* about the tens of thousands of lines of networking code necessary to support every protocol and concentrate on more important client issues."[34] This is the sense in which proxies can act as standardized infrastructure: they allow us to forget. There are other ways of establishing a computer user's location, but proxies can act as sufficient delegates when needed. Representative democracy is in some ways just such a labor-saving device. It's a technique meant to save the population of eligible voters from having to cast votes directly for each new policy of the government. In each case, it is trust that allows people to use proxies as sufficient delegates, and trust that binds people to proxies as faithful stand-ins.[35]

The work that proxies do to make systems function is invisible and easily forgotten by design. There is now a prevailing understanding of what is meant by "infrastructure" that is echoed here, where infrastructure is an expansive category that includes the taken-for-granted conditions and resources that allow the day-to-day operations of the world to take place—including not only the pipes, roads, and cables that act as conduits for information and goods, but the people, paperwork, standards, and protocols that give it all sense and shape.[36] Running a VPN and electing someone to represent your geographical district have very different stakes, and the consequences of "forgetting" about these proxies are also unequal. The project then becomes a matter of remembering and further documenting the consequences of delegating to proxies the power to represent the world. We are surrounded by proxies, but some proxies matter more than others.

Consider the Consumer Price Index (CPI) and its market basket of goods, which acts as a stand-in for consumption habits and their costs. The market basket of goods contains a selection of everyday commodities used to measure and communicate the economic changes that people feel the most, such as inflation. The CPI basket is an inordinately powerful proxy that establishes, among other things, a benchmark for wages and social program funding. It does this by measuring the current and changing costs of everything from food, housing, and medical care to clothing, cars, and education. Determining the composition of the market basket is a precise and arduous task, full of conflict and disagreement.

While fighting inflation in the early twentieth century, statisticians produced relative price indices for a "subset of goods purchased by working-class families" and then calculated the difference between the ratios of expenditure.[37] Although they agreed that the market basket of goods could act as a proxy for consumption habits (and "working class"–ness could act as a stand-in for the health of the nation), they disagreed on what goods should be included in the basket and what year should act as the baseline.[38] While proxification wasn't in question, the specific character of the proxy was. In the ongoing maintenance of the CPI basket, economists and statisticians must decide which products to track and how to measure their costs based on a range of factors, including the place they were purchased and the time of year. They also debate how and when to replace items in the basket in order to reflect the cost of living most accurately. Here's a real question: how can one account for changes in cars every year when changes to the quality of manufacturing are difficult to separate from the clutter of marketing?[39]

When the CPI came under criticism in the 1990s, some economists argued that the difficulty of calculating it was related to the increased complexity of modern life: "a larger fraction of what is produced and consumed in an economy is harder to measure than decades ago when a larger fraction of economic activity consisted of a smaller number of easier to measure items such as hammers and potatoes."[40] This is a claim about the changing quality of American consumerism, but it's also a claim about the limits of proxification; as we graduate from hammers and potatoes to three-dimensional printers and a wide array of complex carbs, we can also see the ways that standardized proxies meet their capacities to be faithful delegates. There are few proxies with either the scope or the influence of the CPI's market basket, but without frameworks for interrogating proxies, we have very few tools at our disposal to question this vital instrument from our vantage point outside the field of economics.

The legal system is also full of proxies. From law school forward, lawyers are trained through moot courts to imagine and simulate the course of argumentation, and mock juries are regularly used to anticipate and predict the results of trials. Perhaps the most common legal proxy is the use of the "reasonable person" standard—what Mayo Moran calls "the common law's most enduring fiction."[41] The reasonable person has a number of siblings, including the "man of business," the "officious bystander," "the reasonable juror properly directed," and the "fair-minded and informed observer," all of whom form a "select group of personalities who inhabit our legal village."[42] The reasonable person is a projection of a proxy: an imagined, rational member of the community who interacts with the world in ways that judges and juries imagine that a reasonable, rational person ought to. In English courts of the late Victorian era, the reasonable person was referred to as "the man on the Clapham Omnibus"—a name that is still

THE POLITICS OF "AS IF"

Proxies are the necessary and practical products of suspended disbelief. To see them this way is to see their usefulness as practical analogies. Institutions need proxies that stand in for real phenomena as if they were the real thing. In the early twentieth century, the German philosopher Hans Vaihinger published *The Philosophy of "As If,"*[73] in which he referred to the most important of these kinds of analogies as "fictions." We willingly accept fictions, says Vaihinger, because the world is otherwise too chaotic and irrational to explain and manage. His general thesis was that many important ideas, around which institutions and disciplines form, are strictly and logically contradictory.[74] Nonetheless, we accept them as true enough because they are *useful* untruths.

Vaihinger didn't think that it was necessary to reject fictions as simple falsehoods—to be a pure skeptic—but rather to understand them as the inescapable artifacts of human thought. He writes, "It must be remembered that the object of the world of ideas as a whole is not the portrayal of reality—this would be an utterly impossible task—but rather to provide an *instrument* for finding our way about more easily in the world."[75] And here, we find a glimpse of the cultural work surrounding proxies. Proxies, as fictions, are instruments that draw their power from repetition and reiteration, through the ways that they form particular habits of use and reference, and through the ways that communities affectively bond to these collective practices of make-believe. We can look to the material and cultural settings of their genesis, circulation, maintenance, contestation, and repair to understand why they persist. In other words, proxies are necessary untruths that nonetheless operate as if they are true "because it is useful for some purpose to do so." We must turn to the politics of "as if" to understand the uses and purposes of suspended disbelief.[76]

Vaihinger singles out the *homme moyen* (the average man) from the nineteenth-century work of Lambert "Adolphe" Jacques Quetelet as an especially important example of a useful statistical fiction, what he calls a "fictitious mean."[77] The nineteenth century was a boom time for the average. As William Stanley Jevons wrote in 1874, the average "enables us to make a hypothetical simplification of a problem, and avoid complexity without committing error."[78] The average man was just one such simplification: a composite, abstract figure that represented the distribution of several attributes of the population of a given country according to a binomial curve (also known as a normal distribution, or "bell curve"), which could then come to serve as the "type" of the nation and "the representative of a society in social science comparable to the center of gravity in physics."[79] The average man, and the calculation of frequencies that came with it, was fundamental to the development of statistical science and state population management—and a building block of eugenics—in the nineteenth century and beyond.[80]

Building on his development of the average man, Quetelet undertook a lifelong study of human traits and activities, leading to the development of "moral statistics" that sought to identify the "propensity" of particular classes of people to, for example, commit crime—a criminology based in race science and statistical averages. Bolstered by the fact that the average man was, for Quetelet, both a national and a racial type, the average man became an instrument to tie a particular quantification of whiteness to a national identity and to criminalize those who were not reflected in its idealizations.

For Quetelet's early contemporaries, it was necessary that the use of the average man was only theoretical, and any suggestion otherwise was met with ridicule.[81] But for Quetelet, the average man was far from a mere fiction.[82] His great innovation was turning the assumptions undergirding normal distribution on their head: instead of thinking of statistical probabilities as the composite product of real phenomena, Quetelet imagined that if a normal distribution curve were a natural law, it could be harnessed in the production of more normal populations. The average man was not just an instrument for thinking through statistical norms, but also an instrument for making normalcy incarnate; as averageness could become a template, decisions about social management could be directed toward maximizing the reproduction of such an ideal. This is borne out in the history of the average man's journey from the domain of probability to the development of a science of populations.

the University of Southern California (USC) cropped from the November 1972 centerfold of *Playboy* magazine.

Chapter 3 examines the institutional setting of digital image processing at USC. I document the environment in which the Lena image could seem like a possible solution to a range of test image problems: the need for a human face, the need for complex images, the need for new images, and to the apparent problem of an overabundance of so-called boring images. Here, I look at how all of these "needs" became cover for importing mainstream, soft-core pornography into the earliest days of networked image transmission, and examine the work that early image engineers were doing at USC on image detection and transmission.

Chapter 4 turns to the late twentieth century, looking in particular at the early 1990s, a time when the graphical World Wide Web was on the horizon. This was also a time when digital image processing was distinguishing itself from the cognate fields of optical engineering and signal processing. Following a feminist media studies approach to this history—one invested in a politics of change and a commitment to reducing and redressing injustice—this chapter looks at moments of resistance to the alienating and often abusive environments of computer science and image engineering, tying conflicts in these environments directly to the visual culture of test images. Together, chapters 3 and 4 argue that the methods of "seeing like an engineer" that produced the Lena image are a product of institutionalized, professional vision, inescapably tied to the practices of decoding and instrumentalizing women's bodies as test data.[96]

These two chapters serve two historical purposes: first, to tell the underexplored story of the earliest days of digital image processing and the attempts to get digitally processed and compressed images onto ARPANET—the direct predecessor of the internet; and second, to write the history of the creation, circulation, and canonization of a *Playboy* centerfold as a test image. I examine how gendered practices shaped image engineering labs and how the very concept of gender was performed and reencoded in image analysis practices and techniques.[97] Methodologically, the chapters draw on an archive of journals, working papers, and gray literature. This includes unofficial reports that documented the work that many engineers, students, and workers did to contest the sexist settings of computer science and engineering throughout the 1980s and 1990s.

Chapter 5 contains a history of the standardized patient program, which began at USC (again!) in the 1960s and transformed, over forty years, into a necessary part of medical accreditation in Canada and the United States. It tells a story in which the surrogate logic of standing in was extended to include human beings as standardized proxies. In the standardized patient program, actors embody the typical symptoms of a disease and trainee physicians diagnose them while honing their bedside manner. Begun as a bridge between the dissection of cadavers and living anatomy class, the standardized patient program functioned as a "living cadaver" lesson.

Through the refinement of the program, standardized patients became a technique for training doctors in diagnosis and the emotional management of patient interaction—techniques intentionally engineered to help physicians avoid malpractice lawsuits. In spite of the fact that actual patients are both vulnerable and unpredictable, standardized patients are meant to be neither—since to be either would threaten the testing scenario they enable. Despite this, it is their shared humanity, the immanent possibility of them becoming patients, that allows them to stand in. As "patients," they act as a gauge, recording the accuracy and affect of their trainee physicians; for medical educators, they act as a consistent test scenario that can be used to compare students. Yet, unlike the kilograms and test images of the previous chapters, standardized patients talk back: they emote, they adjust, they feel pain, they are prejudiced, they mask their own traumas, and they bring with them a lifetime of interactions with the medical establishment.

Whereas the other artifacts examined in this book manifest in things like pieces of metal, paper, and pixels, standardized patients are maintained not only *through* the bodies of workers, but *within* them. Standardized patients reveal a limit for the surrogate logic of proxies, as they chafe at the ability to create predictable and reproducible testing scenarios and show how messy encoding a stand-in can be. But all proxies are messy, and each of the histories included here contains contingent, makeshift, and ritualized forms of labor that workers use to justify and maintain the use of certain materials over others. This labor aims to conceal and suppress the arbitrary nature of scientific and technical decision-making. The messiness of standardized patients simply brings these issues to the foreground. The book concludes with a recurrent theme: the inescapable fact that the naturalization of infrastructure and standards requires a great deal of labor to be successful. The argument, I hold, is that the seams in interwoven technological systems need to undergo constant concealment to appear smooth. By looking through the lenses of artistic appropriation and through critical infrastructure studies as a form of perspectival denaturalization, I offer a methodological detour suggesting how the history of proxies might map another way of surfacing the relationships that hold technologies together.

When issues appear with proxies, it is not in the form of some sudden, catastrophic failure; instead, issues appear *as* issues only when the fabric of communal referencing strains under the pressure of some other social demand. In the stories included here, these demands include the instability of platinum-iridium; the politics of gendered representation; the rigidity of copyright ownership; the legal consequences of bad medical care; and the capacity to speak about one's own pain.

The resolution to these problems will not simply arrive as new and better proxies. These are not merely struggles over the arbitrariness of picking one fixed point over another; they are struggles over the power to pick *any* fixed point, the ability to contest the circumstances of one's work, and the very possibilities of standardization. The power of proxy logic resides in our imaginative capacity to inscribe and realize a vision of the world and to fabricate scenarios where people, places, and things can reside in measurable comparison. The power to determine proxies, therefore, is nothing less than the power to determine the grounds of difference.[98] Who makes that difference, ultimately, is always open to debate.

2 HOW TO CLEAN A KILOGRAM: STANDARDS, DATA HYGIENE, AND THE THEATER OF OBJECTIVITY

THE CONVERSATION ROOM

It is October 22, 2018, and I am sitting in the Conversation Room at the Royal Institution of Great Britain (RI). The RI is a venerated site of public science education made famous through the exhibition of new discoveries and technologies, including Michael Faraday's demonstrations of electromagnetism, Guglielmo Marconi's demonstrations of wireless communication, and Nikola Tesla's spectacular demonstrations of alternating current. In the late nineteenth and early twentieth centuries, public events at the RI were often written up in the press, and represented the site of an idealized middle-class audience, as Carolyn Marvin describes.[1] For over two hundred years, the RI has served as the place where theoretical and scientific insights can become consumable events, and where the public can see how those insights will affect everyday life. It's a theater for sharing emergent scientific ideas with a privileged audience. Tonight, I am here to see Michael de Podesta, an experimental physicist and employee of the National Physical Laboratory (NPL).[2] He is scheduled to give a talk called "The Measure of Science: Redefining the Kilogram."

As the crowd quiets, de Podesta begins with a simple statement on the nature of measurement: "Measurement is an incredible idea; it's a very, very simple idea," he says.[3] There are roughly a hundred people in the Conversation Room. I am here because, for the past six years, I've been studying the management of the metric system and the methods that scientists and

Figure 2.1
Clean, official kilograms under bell jars. Image: R. R. Mulvin.

technicians have developed for maintaining standard kilograms as proxies for the idea of "mass" (figure 2.1). The title of de Podesta's talk refers to the fact that the definition of "mass" is about to change. Mass has been based on one physical artifact—the International Prototype Kilogram (IPK)—but now it will shift to a definition based on a numerical constant. Right now, I don't really think that measurement is very simple. But it's a rare opportunity to see the changing definition of a standard, and the deliberate end-of-life planning of a proxy—one that has persisted for an inordinate 130 years.[4]

Then de Podesta continues:

> Measurement is just this thing where you notice things; you notice that one thing's bigger than another. And, after a while, you stop just noticing one thing's bigger than another, and you do something like you pick up a standard stick and you say, "This is the one I'm going to use to measure whether that plant's bigger than that one": you put it against one and you measure it; and you put it against the other and you measure it. People have being doing this since time immemorial.[5]

And there it is: at the heart of de Podesta's simplification of measurement is a proxy—in this case, a standard stick—that becomes a fixed point through which comparison is made possible and systematic. For de Podesta, and the metrological community more generally, having a communal standard is foundational to the very idea of measurement. When pressed for the pithiest of definitions for what measurement is, de Podesta says, "What is measurement? I've thought rather long and hard, and I've come up with this two-word definition: *quantitative comparison*. Or *quantitative comparison (of an unknown quantity with a standard)*."[6] The shared standards that de Podesta describes—those common benchmarks—are some of the oldest and most pervasive of proxies: stand-ins for ideas like "length" or "mass" that are fixed points, ready-at-hand, and against which other things can be compared.

Finally, he reached his thesis:

> The key to our understanding the world around us is the fact that we can measure. And so the key thing I'd like you to realize is: the only way we know anything about the world is by measuring things. And it's measurement that makes science scientific; it is not mathematics.[7]

How did a presentation in the Conversation Room about kilograms lead to a statement about "the only way we know anything"? We live in an era of unprecedented quantification and comparison. New standards abound to measure and classify people, behaviors, and phenomena. Some of this is engineered for profit, and some for the control of populations—the long hangover of the so-called average man—but all of it relies on the "very simple idea" that things can be compared if there is a shared benchmark. This is an ideological way of seeing the world as revealed through comparison. And whether or not we accept this view as legitimate, we have to reckon with the traction it attains and the manifold ways that benchmarks structure and distort our lives.

Fast-forward a few months: in May 2019, the definition of mass in the metric system was officially revised, and the IPK was replaced with a new standard—a standard that stipulates a recipe for creating or calculating a kilogram under strict conditions, based on the value of the Planck constant.[8] The IPK was replaced by a calculated value meant to provide a new, more stable base for the measurement of mass. But the convoluted replacement process occurred despite the fact that the original system was working just fine. As Terry Quinn, the former head of the International Bureau of Weights and Measures, argues, "There is no evidence of there ever having been a problem with mass measurement or measurement of any other physical quantity whose unit depends on the kilogram that could be attributed to defects in the system."[9]

Prior to this redefinition, the metric system's definition of mass was equivalent to the mass of the IPK—a cylindrical piece of platinum-iridium held in a vault in the suburbs of Paris. By basing the measurement standard of mass on a literal piece of metal, both the standard and the object were defined only with reference to themselves. This meant that without external checks, strict protocols were required to keep the kilogram as static as possible. To do this, the IPK needed to be compared with other, sibling kilograms—*but only after washing and cleaning*. In other words, the common standard that Michael de Podesta identified as foundational to the idea of measurement, relied on the human, manual, and embodied practice of hygiene. This was a curious and ad hoc process that reveals the haphazard, often fluky ways that proxies persist because they have been deemed fixed points.[10]

This chapter examines the IPK through the rituals that made it a viable fixed point. The IPK was born via ritual and returned to ritual forms of maintenance throughout its existence. Ritual, in this instance, is a way of distinguishing some objects as sacred, and it is best understood as a way of construing a protocol or an act so as to invest it with importance and meaning.[11] To understand the ritual labor of making and maintaining proxies, I trace two converging trajectories. The first considers *data hygiene* as a set of techniques and practices for keeping knowledge and information systems orderly. With particular focus on the ways that order is attained through embodied and manual protocols, data hygiene provides a heuristic for seeing the pervasive cultural labor that exists in the maintenance of knowledge infrastructures. The second trajectory follows the study of standards as specific examples of knowledge infrastructures by building on existing studies of standards as sociopolitical artifacts and "recipes for reality."[12]

Bringing these two together, this chapter uses the IPK to illustrate some common features of proxies, as well as the protocols and rituals of maintenance and repair that make standards possible. Proxies are both porous (they absorb their surroundings) and sticky (they pick up pieces and leave traces of wherever they travel), leaving them marked by their institutional management and the communities of use where they circulate. To play off of another use of "fixed point" and to follow Ludwig Wittgenstein: "The axis of reference of our examination must be rotated, but around the *fixed point* of our real need."[13] If we let cultural labor and the politics of embodied performance be our fixed points as we examine the IPK, then we can situate the attempts to mitigate its porousness and stickiness as attempts to render the visible invisible and the detectable undetectable—to erase, in other words, the markings of its history, its emplacement, and its use.

DATA HYGIENE

Everyone has purity rituals: some people wash their hands in sinks, others douse themselves in alcohol-based hand sanitizer, we purge with fire, or we cleanse our bodies with specialized diets. In every case, dirt is shunned through learned, practiced, and embodied techniques.[14] "Dirt is essentially disorder," Mary Douglas writes. "If we shun dirt, it is not because of craven fear, still less dread of holy terror. Nor do our ideas about disease account for the range of our behavior in cleaning or avoiding dirt. Dirt offends against order."[15] Treating dirt as a category instead of as a natural thing allows us to see the social practices behind hygienics, including the fluid and changing boundaries of what constitutes dirt. It also allows us to view and analyze the technologies, protocols, and rituals that bring the category of dirt into being.

If dirt-as-category includes all forms of disorder that need to be cleansed, then it is also consistent with the norms and rituals that pervade in information and knowledge systems for maintaining "clean data," where data scrubbing and cleaning are both "accepted and unexceptional."[16] Anyone working with collections of data, whether large or small, will recognize the importance of maintaining a clean data set. Cleaning in this case involves the removal of extraneous, erroneous, or inconvenient elements from a database to create usable and verifiable data. In financial operations, for instance, data scrubbing or data cleansing might be as simple as rendering the data in the same unit (e.g., gold, silver, or a national currency). This practice of bringing disparate data points into relation, finding common units, and scrubbing unwanted information (to remove disorder) is a manual and repetitive process.[17]

In a database, a valid data set "must look pristine at the end of its processing."[18] To enable comparison and commensuration, certain features of data must be brought into focus (polished), while others are scrubbed clean away. Dirty or unclean data are not bad in and of themselves, but they do represent a threat to the coherence and usability of a data set, technology, or system, where eliminating dirt "is not negative movement, but a positive effort to organize the environment."[19] Data hygiene, then, is a way of describing the cultural work of articulation and disarticulation that bring data into accordance with their intended function by keeping "dirt" at bay.

Laundering

In the early 1970s, anyone following the Watergate scandal could add a new term to their lexicon when they learned that money could be "laundered." Specifically, they learned that President Richard Nixon's former commerce secretary, Maurice Stans, in his new role as finance chairman of the Committee to Reelect the President (CRP for short, CREEP for fun), had laundered illegal campaign donations through a Mexico City bank—donations that eventually went, among other places, to bribe the Watergate burglars and to buy First Lady Pat Nixon a pair of diamond-studded earrings.[20] The term "money laundering" did not appear in a major American newspaper until 1972, and throughout the Watergate scandal it usually appeared in scare quotes. In *All the President's Men* Bob Woodward and Carl Bernstein describe how Martin Dardis, the chief investigator for the state attorney's office in Dade County, Florida, linked the flow of money from a Mexican bank to Nixon's campaign for reelection. "It's called 'laundering,'" Dardis said. "You set up a money chain that makes it impossible to trace the source. The Mafia does it all the time."[21] Dardis was right about what had occurred with the president, and his apparent familiarity with the term "laundering" indicates that both the practice and term predated the scandal, although they had yet to penetrate common parlance.

Why would the Mafia, or a finance committee, want to launder money?[22] In each case, the aim is to clean money (as data) of the traces left by suspect and illegal sources; currency may be fungible, but the records of money's exchange are not. The traces left by money as an exchange medium can form their own record of what Lana Swartz calls a "transactional community": the set of relations that are produced by, through, and around transactions.[23] In the case of ill-gotten or ill-spent money, laundering is an attempt to camouflage this community. We can imagine money launderers like Stans—who kept the records of Nixon's laundered money—bent over their ledgers, deciding what bare minimum of data was necessary to keep for their own records and what needed to be effaced to conceal their transactional community. This is also a laborious memory practice of deciding what sufficient and plausible set of traces could conceal an illegitimate or dubious past.[24] We might also imagine ourselves staring at a spreadsheet of grades, ethnographic research, or bibliographic metadata to carry out the time-consuming practice cleaning data, pecking away at the DELETE and TAB and ENTER keys until order has been restored.

Voiding

In the simplest understanding, hygiene is a necessary and laborious part of making data commensurable and usable. But this assumes that we know and agree on what constitutes the extraneous or erroneous information that needs cleansing. "There is no such thing as absolute dirt," Douglas writes; "it exists in the eye of the beholder."[25] Take, for example, the US Supreme Court's 2018 ruling on the maintenance program for Ohio's voter lists (*Husted v. A. Philip Randolph Institute*). At issue was the purging of voter lists that was done ahead of the November 2016 elections. Registered voters in Ohio who had not voted for six years and who had not returned a postcard that they were sent were removed from the eligible voters' list.[26] Ohio was not alone; across the United States, many jurisdictions were eliminating voters from their rolls in similar ways. The states purging voters claim that such data hygiene is necessary to prevent voter fraud, though it appears to be a naked attempt to suppress votes by sowing confusion at the polls and excluding otherwise eligible voters from participation.

The technique has clear origins in the interpretive flexibility of existing laws, identified and exploited by conservative activists. In 2014, the Heritage Foundation, a conservative think tank, published a fearmongering report titled *A Primer on "Motor Voter": Corrupted Voter Rolls and the Justice Department's Selective Failure to Enforce Federal Mandates.*[27] The primer, written by a lawyer at the Virginia-based Election Law Center, instructs potential activists on how to exploit the 1993 National Voter Registration Act (NVRA), or "Motor Voter Law," to reduce their voter lists. The techniques rely on a selective reading of the NVRA's mandate. The primer states, "The third goal of the law was to impose a minimal obligation on states and local election officials to maintain *clean* voter rolls through the implementation of a *list maintenance program.*"[28] This was a politicized interpretation of what constitutes a "clean" voter roll, which has allowed some states to remove less frequent voters, as opposed to only those who have died or moved away. The Supreme Court ruled 5–4 in favor of Ohio's list maintenance program, upholding the state's interpretation of what could constitute a clean voter roll and paving the way for future voter-suppression techniques.

At the forefront of voter suppression in the United States is an opportunistic interpretation of what constitutes "dirt" in a data set, as well as conflicting norms over hygienic protocols for maintaining clean data. Evident in this case is the fact that the meaning of extraneous data is a contested area; there is a political stake in categorizing list items, and political actors develop highly choreographed rituals to justify these categorizations. The ruling in *Husted* confirmed the legitimacy of a scripted performance of citizenship, in which an unreturned postcard combined with a pattern of nonparticipation could transform a person's name into a political commodity, where deletion became a gain for some and a loss for others—though, we should stipulate, the easy deletion of anyone's name is a loss for everyone with a vested interest in supporting access to electoral democracy.

Sanitizing

In the development of contemporary information professions, as Michelle Murphy writes, office buildings were "machines designed to encourage the buzz of 'information' work inside and to produce a clean, orderly corporate world sealed off from both the polluted outdoors and the dangerous factory floor."[29] Likewise, in the early twentieth century, filing cabinets were promoted and adopted as technologies for compressing documents, managing an overwhelming amount of paperwork, and responding to the corresponding demands for the storage and retrieval of information.[30] Vertical filing cabinets sat on floors, but manufacturers would offer "sanitary legs" that could hold the cabinets several inches above the ground, permitting a cleaner to sweep under them.[31] In the case of the office, then, there are nested understandings of cleanliness that cannot be easily separated: the information in the office needs to be processed to be clean—well organized, easy to recall, and free of extraneous data—but this is possible only if the environment is itself free of contamination.

In homes and sites of leisure, we are also surrounded by familiar hygienic protocols. And just as the office contains nested understandings of cleanliness, homes are also characterized by the interdependence of systems for maintaining order. Ruth Schwartz Cowan documents the changing character of housework as gendered labor and in doing so makes the case against the separation of any one task (e.g., cooking, cleaning, shopping, caring) from any other—since each will rely on the other.[32] In the case of data hygiene, we might focus on those practices that are exerted on domesticated media texts: acts of censoring, cropping, dubbing, bleeping, moderating, flagging, filtering, blocking, and deleting content that allow texts to cross the threshold between the public and the private. The Motion Picture Production Code, the Communications Decency Act (CDA), and social media content moderation policies are all top-down methods of trying to control media content and to enforce normative conceptions of acceptable representation in the public sphere.[33] To view these policies and regulations as practices of data hygiene is to understand them as techniques for managing the contested boundary between clean and dirty, as protocols for performing hygienic labor, and as ways of making a public media experience commensurable within prevailing and hegemonic standards of decency.

For media texts, Raiford Guins draws a distinction between cleaning and sanitizing. The term "sanitizing," according to Guins, refers to the ways that images, films, television programs, and music can be edited to produce, for instance, so-called family-friendly versions. Hearing a song on the radio with the curse words bleeped out is a sanitized version. But sanitizing leaves a censorial gap, Guins argues, and "cleaning" is a further act of repair that not only sanitizes the text but fills the gap left by the initial cut. A clean version of the same profane song would substitute new words to cover up the censorship, and the result would appear to be the original.[34] In both cases, hygienic practices are instrumental in making texts that can circulate according to differential norms of propriety. Sanitized films, as Guins points out, are often promoted as versions that audiences can supposedly trust. If dirt is equal to disorder and dirty content is untrustworthy, then hygiene becomes a way of bringing trustworthy content through an imagined boundary protecting the home.

In American laws and regulations that seek to control content through a threshold between decency and obscenity, children are often the presumed victims of unfettered access to any allegedly dirty content. If we take data hygiene as an umbrella term to refer to a range of hygienic practices meant to identify and manage data, information, and content, then the idea of "out of place" can mean anything from an inadvertent keystroke that infects a data set to a fleeting expletive on live television, to an image that must be vetted before it is posted to a public website. From the office to private media lives, hygiene is not a supplemental practice added to the management of information, media, and data. Instead, it is a constitutive condition for the emergence of information management, as well as the very architecture of private and professional space—internal, external, and the ventilation in between.

>>>

Hygiene is a fundamental technique for delineating social boundaries and norms of orderliness; it is a way that a society constructs techniques for identifying some things as dirt, and the response to that dirt in the form of rituals and practices of cleanliness. But hygiene is also inescapably bound up in power. Hygiene protocols always target specific things, people, and behaviors. And they are carried out by specific bodies. In the United States, voter suppression takes place in a system of structural racism that, in concert with the state's carceral industries, excludes racialized, immiserated, and marginalized populations from the democratic process; databases are often constructed to limit gender identity to a binary value, even when user-facing interfaces present a variety of options;[35] content moderation, on social media platforms, is often specifically targeted at limiting the communication of marginalized and minority users and of sex workers, and is carried out by a precarious and outsourced workforce often denied the bare minimum of workplace safety and mental health protections;[36] custodial labor is disproportionally carried out by racialized and migrant workers; household labor is disproportionally carried out by women. And some users of information systems are considered to be dirtier than others.

As T. L. Cowan has suggested, "digital hygiene" can refer to the ways that we are called upon to maintain clean digital habitats. This might mean keeping an orderly file structure, or maintaining a professional presentation on social media. In either case, the metaphor always concerns the body.[37] Just as bodily hygiene, diet, and behavior have been used to shame individuals and to police bodies that are deemed a threat to public health, the compulsory demand that our digital environments be sanitary falls mostly upon the already-marginalized. Even when we look for exceptions to this rule, we find that hygiene is inescapably bound up in political negotiations over the power to handle information. As Daniela Agostinho and Nanna Bonde Thylstrup show, whistleblowers, leakers, and "truth-tellers" are always "entangled in gendered matrices of control that make possible some truth-telling subjects while foreclosing others."[38] All of this orients the study of data hygienics (1) as a social category for the policing of bodies and behaviors, (2) as relying on the embodied and infrastructural labor of a community of practitioners, and (3) as bound up in a political negotiation over who can handle and manage information in normatively valued ways.

As the remainder of this chapter returns to the role of standards in structuring lifeworlds, it will be tempting to see the protocols for maintaining the IPK as being unbound to the same political negotiations over hygiene protocols. But this couldn't be further from reality: not only is the metric system the foundation of a global infrastructure for the measurement and control of space and time—a European-based standard on which countless other standards rely—but the protocols for cleaning the kilogram demonstrate that, at the most fundamental level, standards and infrastructures are *always* tied into the manual and bodily practices of hygiene and cleanliness that make data socially valuable.

STANDARDS AND THE THEATER OF OBJECTIVITY

Standards are a dominant way of organizing the world. They distribute people, places, and things according to hierarchies, classifications, and categories, and they do so in ways that are putatively pragmatic. Standards are often treated as obdurate pieces of technology that spread over time and space with a force all their own. But in practice, the standardization process is contingent, as people craft standards in specific places and for practical or idealistic reasons, and they maintain them in makeshift ways according to local imperatives.[39]

Over the past several decades, scholars have increasingly turned to the role of standards on two contrasting scales: as technologies that condition everyday existence, and as technologies that span the globe, enabling systems of trade, circulation, and management.[40] But on each scale, standards are made to get things done—for some people, at some times. As we are constantly reminded, standards are also meant to disappear from view, to sink to the level of infrastructure, to go without notice, and to become second nature—they are meant to help us to forget.[41] As Elizabeth Cullen Dunn says, "A standard without an appropriate infrastructure cannot be put into force without major upheavals in the physical environment and the social organization of production."[42]

To employ standards requires an initial investment in a single decision about choosing fixed points.[43] These points are both arbitrary and precise:

even when based on supposedly invariant, natural phenomena, they are the product of human community, human measurement, and a collective agreement to accept them as fixed and communal.[44] For a familiar case of arbitrary precision, take the case of progress marking in football, in which officials manually place the game ball on a line at the end of each play by basing the measurement on their split-second judgment of a tackle. They determine, to the best of their judgment, exactly where the football stopped moving forward. This location becomes the fixed point for starting the next play. In instances where the ball is very close to the threshold for a first down, officials will bring out a ten-yard chain to measure if the ball progressed far enough.[45] The entire scenario provokes necessary reflection: why employ a precise measurement instrument like the chain when the first judgment was imprecise and reflexive? This is the logic of arbitrary precision: decisions must be made to situate fixed points in certain positions, and those fixed, certain positions become crucial to further judgments. Protocols like the referee's locating of the football are part of the rituals that go into making systems appear objective; they are part and parcel of a theater of objectivity. In another sense, fixed points work as the "fictions" that make standardization possible.[46]

A standard, as Lawrence Busch argues, is a recipe for producing and reproducing a narrowly defined phenomenon. A meter stick stands in for a fraction of the Earth's meridian and is easily reproduced in metal, wood, or string whenever a meter is needed.[47] As a proxy for both a specified length and for the process that created it, the meter stick eliminates the need to remeasure the meridian and protects against the variations in other ready-at-hand standards, like the difference in length of various people's outstretched arms.[48] This is the promise of proxies: they allow us to forget. Because scientific and technical proxies are the embodiment of choices about imagining the world within testing environments, like those of the lab, the workshop, the courtroom, and the office, they are important materializations of a hypothesis about the world, forged into small, manageable, usable, workable, exchangeable, and reproducible chunks. They include prototypes, working objects, measurement apparatuses, lab samples, and test and training data—and are supported and buttressed by a whole ecology of objects, manuals, and memos for referencing, checking, and maintaining those materials. To study the material life of proxies is to study the process of capturing the world in usable chunks, to interrogate the choice of some proxies over others, and to investigate the contextual characteristics of maintaining a material apparatus as the basis of a knowledge infrastructure.

Standards are primarily intended to make things run smoothly for interested parties by reducing local differences and increasing interoperability. As Laura DeNardis argues, interoperability is the growing concern of systems that are built to share resources, standardize parts, and exchange information efficiently.[49] There are clear exceptions in which standards actually introduce friction; for instance, proprietary industrial standards that are supposed to make it harder for new entrants to join a market or prevent users from opening the black-boxed content of their media.[50] But even in these cases, we can say that standards are crafted in and through power: they work for certain people in certain ways—industrial standards might protect the market share of existing operators or the financial benefits of existing copyright holders, while state identification standards might limit who can drive, vote, or have their gender identity recognized.[51] Studying the proper functioning of a standard then becomes about identifying for whom it works and under what conditions. If we stipulate this premise, then we can study standards as specifically situated and insistently normative statements about how the world *should* operate—statements that unavoidably embody and codify ethical statements about who or what is a priority.[52]

The smooth functioning of an integrated economy or a system of centralized governance requires the commensurability of people, places, and things to rank, order, and measure. Complexity requires comparison, and standards create and enforce categories that become means of control.[53] Yet, as Geoffrey Bowker and Susan Leigh Star argue, belonging to a category can either be a privilege or an affliction, a condition of possibility or a means of oppression—or something in between.[54] The ambivalence of standards, norms, and categories can compel the abandonment of radical difference in exchange for recognition in an institution. It is a Whiggish view of history as the progress of rights, which treats the recognition of one's difference in a standard or norm as the surest way to economic and political enfranchisement.[55] The danger of recognition in a standard is that it will crystallize an identity or a category that may otherwise be fluid, uncertain, or undecided.

In the eighteenth and nineteenth centuries, standardization was often imagined to be a pathway to interjurisdictional (if not global) harmony. What we now call interoperability was once a test of a country or an empire's capacity to control the basic terms of measurement and exchange, while basic units of measurement for length, weight, and coinage are often regarded as the earliest form of contemporary standardization. By demonstrating the possibility of standardization in these domains, the legitimacy of standardization was later imported into other disciplines and domains.[56] As Kathryn Olesko states, "The techniques and instruments that produced more accurate weights and measures (which made social interactions as well as commercial transactions more exact) migrated early in the century to the sciences where they formed the nucleus of exact experimental practice."[57] Going further, James C. Scott, in his analysis of the modern nation-state, traces its emergence to the provision of basic units and the standardization of measurement. Standards, Scott argues, are "transformative state simplifications."[58] As such, standards appear as primary and ideal instruments of management, and encapsulations of the very possibility of manageability. By using an arbitrary but precise measurement system, states could manage the much-less-precise and much-more-fluid components of their population.

The crafters of the metric system, operating at the height of the Enlightenment and in the throes of the French Revolution, imagined that a new political unity needed an apparently democratic system of units based on a measurement of the Earth's circumference. In a contemporary setting where standards are often equated with state violence, blunt instruments like standardized testing, and the eradication of difference, it is easy to forget that in the eighteenth and nineteenth centuries, standards were imagined as tools for maintaining fairness and justice.[59] Sandford Fleming, the

Scottish-Canadian engineer who proposed a system of time zones (what he called "Cosmic Time"), believed that the implementation of such a system could bring about global, political, and economic harmony.[60] Max Planck went one step further, suggesting that a system of units based on the fixed constants of nature could be a standard shared for all times, by all peoples, and potentially even by extraterrestrials.[61]

The critical and cultural study of standardization is concerned with the ways that quotidian life and practices of making do are structured by standards and infrastructures that are themselves massive, sometimes violent, technologies rendering things interoperable at the cost of heterogeneity and uncertainty. By approaching proxies as leveraged simulations of the world and treating the standardization process as a cultural, performative, and representational practice, we reverse the relationship between the macro scale of standards and the micro scale of lived experience; instead, we can analyze the ways that local contingency enters the standardization process through the representation of the world in usable ways and the embodied labor of animating proxies. Through the cultural lives of proxies, we can trace histories of the cultural interiors of standardization. The history of proxies not only shows how standards are made to contain messy reality, but also what happens when that messiness inevitably leaks out and new realities seep in.

CLEANING A KILOGRAM

To clean a kilogram, you will need the following supplies:

For cleaning by hand
    1) 1 piece of chamois leather
    2) A 500 ml mixture of equal parts ethanol and ether

For solvent washing
    1) 1-L Pyrex flask containing bidistilled water
    2) 1 tube with a 2-mm spout
    3) 1 bowl for collecting condensed water
    4) 1 tripod that can spin on its vertical axis and extend vertically

5) 1 platinum-iridium disk

6) Filter paper

Time: about 50 minutes

About 6 days before you want to clean your kilogram, begin soaking the chamois leather in the ether-ethanol mixture for 48 hours. Wring out the leather—this helps remove impurities, and you definitely don't want impurities on your kilogram. Repeat this stage two more times by soaking the chamois leather in a fresh bath of ether and ethanol and wringing it out each time.

When your chamois is finally free of impurities, you can begin to clean your kilogram. A kilogram needs to be rubbed with approximately 10 kPa of pressure (figure 2.2). If you do not have a Pascal gauge handy to measure kilopascals,

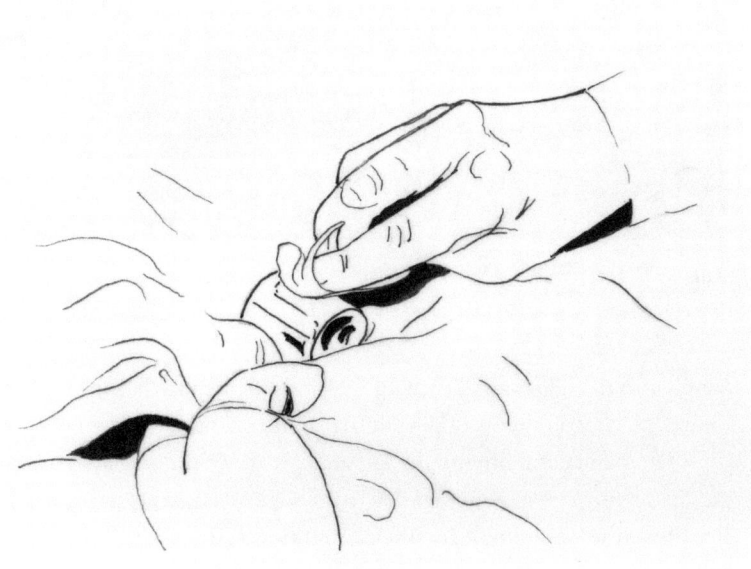

Figure 2.2

Cleaning a kilogram by hand. In your less-dominant hand, cushion the kilogram in a length of chamois leather. With your dominant hand, take a corner of the chamois, wrap your index finger in it, and clean until you reach the handsome, but not specular, surface you desire. Image: R. R. Mulvin.

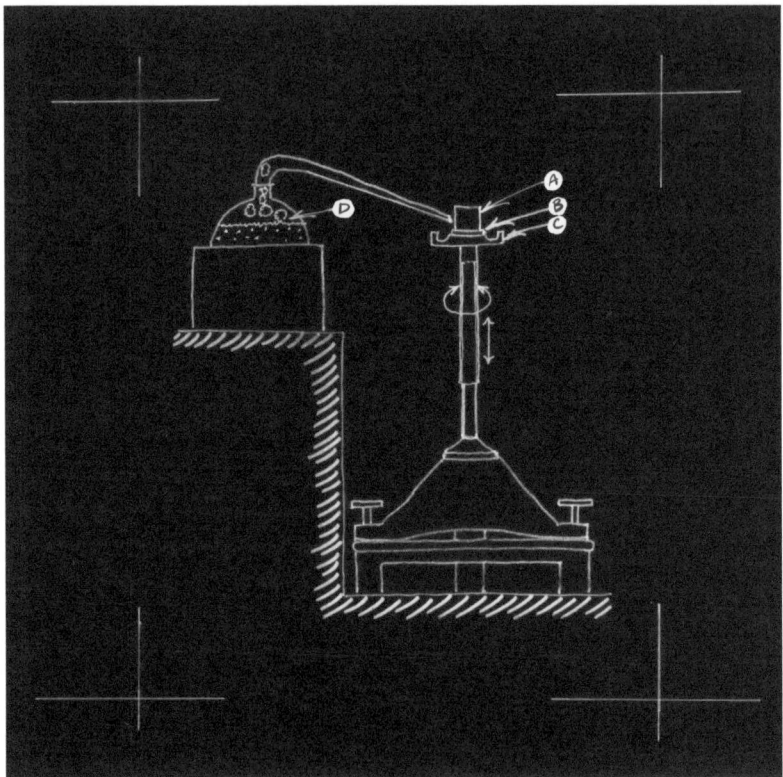

Figure 2.3
The steam-cleaning apparatus for washing your kilogram with solvent. Image: R. R. Mulvin.

simply rub the kilogram fairly hard by hand. Remember that you are trying to return the kilogram to its original luster: "rather handsome, but not specular."

After manual rubbing, your kilogram is ready for solvent washing with steam. Figure 2.3 shows you what your solvent cleaning setup should look like. Place the kilogram (A) on the platinum-iridium disk (B). The disk should fit comfortably on top of the tripod (C). Fill your Pyrex flask (D) about three-quarters with doubly distilled water. Run the tube from the flask to the top of the tripod, pointing the opening at the kilogram but keeping it about 5 mm from the surface. Heat the water with an electric mantle, operating at 350 W.

As the water boils, steam will start to come out of the tube. It should first be pointed directly at the kilogram's uppermost edge. Rotate the tripod on its vertical axis, making sure to steam all 360 degrees of the cylinder. It's like a vertical rotisserie! Start moving the kilogram upward while continuing to rotate it. Blast the kilogram with steam for 15–20 minutes.

Most of the steam should run off the kilogram and into the bowl. It is normal to have a few drops of condensed steam on the kilogram's surface. For these stragglers, take a corner of the filter paper and absorb each drop individually using the paper's capillary action. If you don't have filter paper, you can also blow the drops away with a jet of clean gas.

Next, flip the kilogram on to its bottom so that you can clean the part that was resting on the disk. Repeat the process of steam cleaning the kilogram—rotating it and raising it—so that the cylindrical surface gets a second cleaning. (Of course, you'll want to make sure that the disk under the kilogram was cleaned in advance using the same technique. But you knew that!)

Return the kilogram to its resting place under a glass bell jar (as shown in figure 2.1). If you are accustomed to using a chemical desiccant to dry the air inside the bell jar, it's really not necessary (but there is comfort in tradition). Congratulations—your kilogram is clean![62]

THE METRIC SYSTEM: MEASUREMENTS MAINTAINED THROUGH HYGIENE

For proxies to work, they have to act as fixed points and have to be taken for granted. For this to be possible, they must be orderly. Because, as artifacts, they soak up their surroundings, leak, and decay, all proxies need to be cleansed or risk falling into permanent disrepair. This section considers one practice of hygiene through an analysis of the IPK and the protocols for keeping this small piece of metal clean—a process that extended its life as a proxy and as a component of the metric system. Until recently, the IPK was the last remaining physical artifact that formed the basis of a fundamental unit in the metric system. Because fundamental units are primary building blocks in standardization, the kilogram was a special case: a physical artifact meant to serve as a shared and immutable reference point. Far from an obscure piece of bureaucratic triviality, the standard of mass is inseparable from everyday life and the calibration of the world around us.

The IPK is a 39-millimeter-tall, platinum-iridium cylinder housed in three nested, vacuum-sealed, glass bell jars and locked in a vault on the outskirts of Paris.[63] Access to the IPK requires five separate keys, kept by three separate individuals.[64] If Hollywood films are to be believed, there are more keys involved in accessing the IPK than launching a nuclear missile from a submarine (which movies have shown me only need two keys).[65] That much security is necessary to guarantee that this small piece of metal is safe and uncontaminated.[66] Until 2019, the IPK—represented more elegantly by its symbol, $\mathfrak{K}$—was the basis of the unit of mass in the International System of Units. The International System of Units, abbreviated SI, is more commonly known as the "metric system," though the equation of the two is misleading. When I write $\mathfrak{K}$, please imagine that I have written "IPK."

$\mathfrak{K}$ was afforded its own special symbol because it is a unique object; though it has siblings and descendants, it is special among its kind. $\mathfrak{K}$ gets its own symbol because of its use in mathematical calculations. Like π (pi) or Ω (ohm), $\mathfrak{K}$ is a fixed value used in the calculation of other values. $\mathfrak{K}$, for instance, is used in the calculation of Ω. Unlike π or Ω, however, $\mathfrak{K}$ previously did not refer to an abstract figure, but rather to a physical artifact. The most important aspect of $\mathfrak{K}$ was that all kilograms referred back to it.

Like many proxies, $\mathfrak{K}$ represented a kind of idealism. As part of the original metric system, it was derived from a scientific endeavor situated in the depths of the Enlightenment and was based on French Republican ideals, which stated that more democratic measurement standards could be based on the so-called invariants of nature instead of the arbitrary whims of kings, lords, or merchants. In this case, the invariant was a fraction of the Earth's circumference (a meridian). While not all proxies are this idealistic, the high-minded surrogate logic behind $\mathfrak{K}$ hinged on picking a piece of metal that could transcend the authority of monarchs.[67]

The choice of a proxy is the choice of a fixed point against which subsequent objects will be judged. Lying beneath the choice to base a measurement system on the invariants of nature is a searching desire for stability in a chaotic world, in which standardization represents a human-built attempt at creating order. By divorcing measurement standards from the whims of monarchs and their emissaries, the designers of the metric system were declaring that the scientific community was the proper locus of power to name proxies.

The management of the metric system is overseen by a three-part bureaucracy, created by the 1875 Metre Convention (also known as the "Treaty of the Meter"): the Conférence Générale des Poids et Mesures (CGPM; General Conference on Weights and Measures), which acts like the UN General Assembly of all signatories to the Convention and meets every four years; the Comité International des Poids et Mesures (CIPM), which is kind of like the UN Security Council to the CGPM, meeting every year and guiding decisions about measurement standards; and the Bureau International des Poids et Mesures (BIPM), which encompasses all of the physical infrastructure of the metric system and its headquarters. The BIPM includes a set of laboratories and offices in Sèvres, in the suburbs of Paris. Together, these different groups and institutions are responsible for guiding changes to the future of metrology and providing protocols to national laboratories and methods for establishing equivalences between national measurement standards.

Despite its origins in a French Republican project aimed at creating democratic standards, the IPK was a rarefied object. Although local measurement standards for length were once embedded in city gates, making them available for traveling craftspeople trying to calibrate their tools, the IPK was, in its lifetime, exceptionally difficult to access. This is a deep but functional irony of proxies, as they need to be both easily realizable and secured against interference. $\mathfrak{K}$ manifests this irony in a very obvious way: most of the world engages with the metric measurement of mass on a daily basis. (Even the pound unit of weight in the US customary system is actually defined as a fraction of the kilogram.) Yet a tiny number of people have ever interacted with $\mathfrak{K}$. It's like the pope, but even more secretive. Each measurement of mass that relied on the soundness of the metric system relied on the fact that $\mathfrak{K}$ was safely kept where almost no one could see it. Yet when it is withdrawn from its safe enclosure to be cleaned, it transitions from the secure basis of being a worldwide measurement system to become a working object and a profane artifact of the laboratory; *hygiene* is what connected the kilogram's idealism with its realization as a scientific instrument.

Sporting the French designation *nettoyage et lavage* (cleaning and washing), the hygiene protocols for ℜ were a response to the object's specific history—crafted in the nineteenth century, locked in a vault—and its position atop a referential hierarchy. ℜ is a privileged case in the history of measurement standards that nonetheless demonstrates many of the basic features that proxies share. It was selected and agreed upon by committee and convention, and it is invested with self-referential importance, such that replacing it required a wholesale rewriting of the metric system. The history of ℜ, then, is both a general history of the manual maintenance of proxies and a history specific to the material conditions and constraints of this one specific piece of metal.

THE HIERARCHY OF KILOGRAMS: TRACEABILITY

Measurement science uses the relationships between objects and institutions to maintain the authority of standards. The referential relationship between ℜ and other kilograms is called traceability. Traceability is a term used in a wide range of disciplines, operations, and practices. Genealogy uses a kind of traceability, as does forensics. In measurement science, traceability refers to "a property of a measurement result whereby the result can be related to a reference through a documented unbroken chain of calibrations, each contributing to the measurement uncertainty."[68] There are two things to note. First, this kind of traceability deals in a logic of material presence. The unbroken chain of calibration means that each kilogram in the chain of kilograms must have been compared, side by side, to a kilogram one level above it in the hierarchy. Second, each step away from ℜ adds to the uncertainty of each kilogram's precise mass. Traceability is the combination of calibration and citation: a measurement operation with a history. As one introductory textbook puts it: "To say that a package of butter weighs a pound means that it has been connected by some long and complicated series of comparisons to The Kilogram in Paris, and weighs 0.4539237 times as much."[69]

In practice, this means that local inspectors possess a kind of working standard that is checked against, for example, county, state, or national prototypes so that they can verify a scale in a grocery store—a practical mass measurement instrument—with the knowledge that the standard is traceable to ℜ (figure 2.4). Most national laboratories have prototypes that are traceable to ℜ; these are referred to as "secondary standards," which are calibrated through direct comparison with each other or ℜ, the primary standard.[70] Traceability is the capacity of a system to account for its connections: in this case, to quantify, document, and disseminate the difference between two individual pieces of metal. But not just any traceable difference will suffice. The kilograms involved in the traceable hierarchy of mass standards need to have the lowest possible level of uncertainty.

Figure 2.4

Traceability in the wild—in this case, proof that a scale at a New Jersey branch of a popular chain of grocery stores was recently inspected. Similar stickers are seen on any scale that an employee operates (deli, butchery, fish, and so on), including the scales used by cashiers. This interdependent network of inspected scales guarantees that the measurable weight of goods is consistent throughout a given store, within a particular state or nation, and that all of them are ultimately traceable to ℜ.

THE PROTOTYPE KILOGRAM: A SYSTEM OF FIXED POINTS

The metric system is the most wide-reaching set of measurement standards in history. And among modern standards, it is also one of the longest lasting, having been in force in some form since 1799. Many of the protocols surrounding the metric system established the means through which international standardization efforts could take place. Tellingly, the 1798–1799 Congress on Definitive Metric Standards, organized by the French government, was likely the first international scientific conference of any kind.[71]

Prior to the creation of the metric system in the late eighteenth century, the Ancien Régime in France believed that the flourishing of the British economy was due to their standard measurement system, and the lack of such a system in France prevented them from participating in the kinds of frictionless trade possible in Britain.[72] A uniform system of measurements, the monarchy's chief minister of the nation believed, could solve the national food crisis, so he recommended declaring the local Parisian set of measurements as the new national standards. As Ken Alder writes, "A modern nation needed a standard, *any standard,* and the surest course of action would be to declare the units used in Paris the national units."[73] Instead, as scientists and bureaucrats worked on a new system of measurement—what would become the metric system—revolution broke out in France and standardized units, universal decimalization, and fixed points based on the Earth were each held up by Republican figures as symbols of the egalitarian possibilities of nonauthoritarian rule. The French Academy of Science wanted a universal measurement based on the "invariants of nature." This was an attempt to erect a standard that could be enacted—that is, performed—neutrally, objectively, and beyond the control of the monarchy.

Here, a subtle shift was made to align scientific objectivity with political neutrality and egalitarianism. Hence, the meter was based on a fraction of the Earth's quarter-meridian: "They vowed to choose a set of measures which would 'encompass nothing that was arbitrary, nor to the particular advantages of any people on the planet;'" and a set of measures that could establish "a uniform language for the objects of daily economic life."[74]

Despite the language of the proposal—the insistence that the measurement system would "encompass nothing that was arbitrary"—there is no natural or guaranteed alliance between something called a "meter" and the Earth.[75] Indeed, as later calculations would show, even the measurements of the meridian were erroneous and based on inaccurate, though no less precise, conjecture about the Earth's shape.

In the metric system, the length and shape of a meridian constituted a way of encoding a necessary fiction as a starting point for a new measurement system. The fiction itself was also a heavily encoded signifier, reflecting the social, political, and economic idealism of a nation on the brink of revolution and promising that citizens would be able to share information and conduct business through a standard that was independent of the sovereign. As an outgrowth of the political upheavals of post-Revolutionary France and Enlightenment ideals, the early years of the metric system were tumultuous—and imposition of the new system was often met with violent pushback. But its designers persisted.

In 1799, a prototype meter and kilogram, called "Le Mètre des Archives" and the "Kilogramme des Archives," were produced using provisional measurements of the meridian. The Kilogramme des Archives was based on three interrelated, fixed points: the length of the meter, itself based on a fraction of a terrestrial meridian, and the mass of distilled water at its melting point. Here, 1 cubic centimeter of pure water equaled 1 gram and, to make the prototype more practical (instead of tiny), its mass was 1,000 grams. Yet once the meridional measurements were shown to be inaccurate, the inaccuracy was ignored. That is to say, the *precision* of the inaccurate measurements and the precision of the prototype's construction were far more valuable for the social and political work that the new measurement units would perform, as fixed points, than the accuracy of their referential measurements.

As Scott notes, three factors in particular led to the metric system's ascendance:

First, the growth of market exchange encouraged uniformity in measures. Second, both popular sentiment and Enlightenment philosophy favored a single standard throughout France. Finally, the Revolution and especially Napoleonic state building actually enforced the metric system in France and the empire.[76]

While Napoleonic conquest within Europe is often credited with the success of the metric system, recent research has highlighted the role of industrial standardization. Beyond mere market exchanges, international expositions became major trading sites and stages for the benefits of standardization.[77] For nearly a hundred years, the metric system ebbed and flowed in its adoption and enforcement. Despite being illegal in England (it was legalized in 1896) and the United States (1866) for much of the nineteenth century, the metric system grew into a singularly powerful and deeply entrenched standard through both industrial adoption and military imperialism.

The true moment of ascendence for the metric system came in 1875, with the signing of the Metre Convention. The conference to sign the Metre Convention took place in Quai d'Orsay, in Paris, and lasted from March 1 until May 20. Over those seven weeks, attendees planned the creation and maintenance of new physical standards for the meter and the kilogram.[78] This meant deciding on the design and selection of new physical artifacts, as well as the establishment of a new bureaucracy to oversee the system. At the 1875 conference, attendees considered whether the Mètre des Archives and Kilogramme des Archives would serve as the bases of the new standards, as the intervening decades had shown these standards to be far from true derivations of the "invariants of nature" on which they were supposedly based. They decided to retain these prototypes as primary models and to produce new meters and kilograms as facsimiles. The choice of an original fixed point, the precise-if-inaccurate measurement of a terrestrial meridian had, it turned out, locked in a system of measurement whose usefulness superseded the drawbacks of its inaccuracies. At each stage of renewal and revision in the metric system, the original, arbitrary decision to base the system on meridional measurements was reaffirmed and resanctified.

A renowned British assayer and refiner, Johnson & Matthey, was chosen to produce thirty prototype meters and forty prototype kilograms to serve as national reference standards for signatories to the Metre Convention.[79] The first three kilogram prototypes were produced together and, as such, 𝔎 is one of three "identical" artifacts, chosen for having the smallest measurable difference to the Kilogramme des Archives. The word "identical" is tentatively used here because 𝔎 was vested with its position atop the hierarchy of mass measurements precisely because it is *not* identical to any other object. When it came time to compare the new prototype kilogram with the Kilogramme des Archives, the measurements were carried out in the grand hall of the Paris Observatory (among the greatest symbols of Enlightenment science), which was located directly on the same meridian that was used to formulate the metric system.[80]

I highlight these rituals because they do nothing to make measurement science more accurate. But they clearly perform a cultural function: through the theatrical performance of objectivity and the rituals of commensuration that surround the manufacture and consecration of new proxies, we glimpse the ways that institutions shore up the arbitrariness of human decision-making. Here, the choice of one tiny piece of metal over others in order to stand as a sacred and self-defining unit was explicitly drawn together with the measurement of the Earth, the legitimacy of academic science, the larger political movements of Republican France, and the institutional markers of seriousness, prestige, and history.

THE RITUAL BURIAL

𝔎 was joined by six other check-standard kilograms, called *témoins* (witnesses). The wording here is important. The meaning of the French words *témoin* or *témoinage* (witnessing) is comparable to their counterparts in English. The witness kilograms are meant to attest to changes in 𝔎. They do not do this through eyewitness accounts or oral testimony; rather, they attest to changes in 𝔎 by being brought into measurable comparison. The only knowledge that we have of 𝔎's absolute mass comes from comparisons conducted with a mass comparator, which is like a kind of scale that *only* measures differences between two objects. The witnesses' bona fides are established through a shared history and through their identity as kilograms: they were produced from the same metallic alloy from the same refining company, some at the same time as 𝕽 and some in the years afterward; and they are housed in the same environment, in the same vault, with the same air and the same exposure to light.

According to the BIPM, any factor that is shared by 𝕽 and its witnesses can be ruled out as a factor contributing to changes in the mass of 𝕽.[81] That is to say, the officials at the BIPM believe that anything that would bear on 𝕽 would also bear on its witnesses. It is afforded more care and attention than most measurement tools, but the use of *témoins* and the attempt to control for 𝕽's historical specificity—to identify what is special to it that might explain some change in mass—highlight a facet of commensuration in all measurement operations: measurement is a process of developing conditions for identifying difference.

In 1889, fourteen years after the signing of the Metre Convention, it came time to vest the prototypes and the *témoins* with the imprimatur of the institution at the first meeting of the CGPM. During the *troisième séance* (third session) of the General Conference, the attendees entombed 𝕽 along with 𝔐, the prototype meter. The two would serve as the basis of length and mass measurements in the metric system for decades to come, but first it needed to be sanctioned by the attendees of the conference. In the early afternoon of September 28, 1889, the attendees of the CGPM gathered to enclose the new prototype meter and kilogram. The minutes from the session describe the process as "*enfermer,*" which translates in a number of ways: "to enclose," "to lock up," or "to sequester." Peter Galison translates this act as "burial" and "interment," which might exaggerate the inaccessibility of the prototypes to future access, but usefully highlights the ritual nature of the event.[82]

Eight objects were enclosed that day: 𝕽 and 𝔐 were each joined by two *témoins*, a thermometer, and a copy of the report describing the ritual. This network of objects created a circuit of referentiality based in a documentary regime of verification, with each piece pointing back to another: the report signifying the consecration of the objects, the thermometer attesting to the fact that the objects are kept at a stable temperature (that of melting ice), the *témoins* as corroborators of the prototypes, and the prototypes as embodiments of the measurement system's authority.[83] The entire vault was locked by three keys, behind an outer door locked by an additional two keys, and the keys were given to three separate officials.[84] This process, consisting of burying objects in front of an audience, with reports of their sanctification, and locking those things away with keys to make them transcend their profane origins, attests to the ritual, manual, and documentary conditions that enable, create, and maintain proxies at the basis of standards. Like Jacques Derrida's analysis of the signatures on the Declaration of Independence, where "the signature invents the signer," the Prototype Meter and Prototype Kilogram had no authority *until* they were hidden from view, locked in the dark along with the documents declaring their power.[85] Likewise, the authority of the documents that are buried alongside the meter and kilogram is performed by the copresence of the consecrated objects. One does not predate or include the other; they work together to constitute a verifiable institution.

The ritual burial of 𝔎 and 𝔐 began a new era in the metric system. There was now widespread international agreement among a plurality of nations to assign foreign (Parisian) artifacts as the basis of their measurement systems. Even in the United States—a country famous for its refusal to use the metric system in civic measurements—the Metre Convention changed the meaning of standard units. The United States was an original signatory to the convention, and in 1893 it changed the definition of its customary units (feet, inches, pounds, ounces) to be based on fractions of the metric system. The allure of international interoperability was too powerful to resist. But the 1889 ritual gets short shrift in the history and philosophy of measurement.

The focus instead tends to fall on the definitions of the meter and kilogram that followed after those artifacts were sealed into the vault. At the third CGPM in 1901, the mass standard was defined explicitly:

> The kilogram is the unit of mass; it is equal to the mass of the international prototype kilogram.[86]

This declaration meant that the kilogram, as a unit, is only ever the mass of 𝔎, and the mass of 𝔎 is always equal to 1 kilogram. By this definition, 𝔎 could not include a measurement error, as it contains no uncertainty.[87]

This is the only way that it can serve as a stable ground for determining difference in other prototypes and, for that matter, any measurable difference of mass.[88] This assertion of $\mathfrak{K}$'s lack of inherent error is strictly conventional. A "kilogram" had no natural meaning except that which is assigned by decree to one particular piece of platinum-iridium. The original Kilogramme des Archives, which $\mathfrak{K}$ was meant to reproduce, was based on both the measurement of the meter and the density of water. The meter, in turn, was based on a fraction of painstaking (if inaccurate) measurements of the Earth's meridian. Objects can seal in provisional, mistaken, or hurried knowledge and assumptions about nature, behavior, and political process. The mass standard, then, is at once a rich index of a political process and a flawed index of a natural attribute of the Earth.

BAD HYGIENE

The circular definitions of the metric system's mass and length standards have produced no shortage of philosophical consideration.[89] Ludwig Wittgenstein turned to the International Prototype Meter to prove a point about the conventionality of paradigmatic thinking. He writes by way of example in *Philosophical Investigations*: "There is one thing of which one can say neither that it is one metre long, nor that it is not one metre long, and that is the standard metre in Paris."[90] What Wittgenstein is arguing is that there are certain concepts (or, in this case, objects) that are paradigmatic and self-sufficient, such that they create the standard by which other objects of that kind will be understood, measured, classified, and otherwise ordered. Some have dismissed Wittgenstein as fundamentally misunderstanding or misstating the function of measurement but he, and likeminded philosophers, expose both the conventionality of measurement and an aporia at the center of standards.[91] It is this kind of paradigmatic thinking that Michael de Podesta invoked when he began his lecture at the RI with his definition of measurement.

An object cannot be compared with itself—we need methods for separating objects from each other *and* from themselves. As Natalie Melas describes, comparison implies both a *comparator* and a *perspective*, as all comparison is situated by who or what is doing the comparison and from what position.

In comparative literature, as Melas argues, the Western canon provided the background against which other literatures could be compared (and, accordingly, shapes the inclusions and exclusions of what is considered serious literature). In mass metrology, the kilogram served that purpose.[92] As Karen Barad states, "there is something fundamental about the nature of measurement interactions such that, given a particular measuring apparatus, certain properties *become determinate*, while others are specifically excluded."[93] In the case of mass measurements in the metric system, $\mathfrak{K}$ appears to create the possibility for a determinate mass property in its traceable descendants; through traceability to $\mathfrak{K}$, other kilograms are granted the property of being kilograms, and gain the seriousness of measurability.

But with no ground of its own, $\mathfrak{K}$ cannot resort to traceability for its own determination. Instead, as Barad writes, "which properties become determinate is not governed by the desires or will of the experimenter but rather by the specificity of the experimental apparatus."[94] $\mathfrak{K}$ gained its status as a kilogram through a piece of circular logic embedded within the written documentation for the metric system. $\mathfrak{K}$'s statutory status as a self-sufficient measurement standard provoked a couple of uncomfortable questions: *Could* a decree be the grounds for determining a measurable property? Is the decree part *or* parcel of the measuring apparatus? Changes that the BIPM made to the definition of mass in subsequent years, short of answering these questions, show how the metrology community has, at least, struggled with their answers.

In its 1921 definition for mass, and, with it, the kilogram, the BIPM tried to obviate the need to determine the true mass of $\mathfrak{K}$. The self-sufficient definition vested the act of determining the kilogram's mass in the institutions of the metric system—in its conventions, practices, and protocols, but also in its vaults, bell jars, and keys. But the need to compare $\mathfrak{K}$ to its *témoins* created friction. By the middle of the twentieth century, BIPM officials were aware of the mounting problem of $\mathfrak{K}$'s poor hygiene. Earlier in this chapter, I stated that $\mathfrak{K}$, by definition, cannot contain uncertainty—it is, at all times, equal to 1 kilogram. This is, in practice, true. However, the kilogram does have an absolute uncertainty—a rate or amount by which its mass is increasing or decreasing—which could, in theory, be estimated through comparison to its *témoins,* though never fully determined (or put more simply: we cannot be certain about its uncertainty). However, there is a more fundamental uncertainty that has prevented any attempt to estimate how much ℜ has changed over time, and that is a form of uncertainty that results from an untimely act of cleaning.

When the assayer and refiner Johnson & Matthey manufactured ℜ and the other forty original kilograms, the committee responsible for overseeing the production specified at which stages, and in what ways, the pieces of platinum-iridium were to be washed and cleaned.[95] At times, this meant cleaning with some combination of steam, alcohol, or distilled water. The final step, when the manufacturing was complete, was a steam bath.[96] (The apparatus used to clean kilograms from 1882 to 1889 can be seen in figure 2.5.) The kilograms were then left to dry under a bell jar in the presence of a desiccant, anhydrous potassium hydroxide.

Great care and meticulous specifications were put into the directions for precise hygienic treatment of the kilogram prototypes, and yet no directions were provided for the upkeep, maintenance, and ongoing hygiene of the prototypes *after* being manufactured. It is important to remember that in the mid-to-late nineteenth century, the possibility of a truly functional metric system was an unprecedentedly large standardization project. Although most countries, municipalities, and merchants had experience with their own measurement standards, less thought had ever been paid to how a material proxy would be maintained for a longer period and in a verifiable way.

Hence, there was no mention of cleaning kilograms again until 1939. At this point, Albert Bonhoure investigated the effect of cleaning on the mass of platinum-iridium kilograms. He used a chamois leather cloth, soaked first in ethanol and then in redistilled gasoline, to rub all the surfaces of the cylinder. Although World War II interrupted his investigations, Bonhoure cleaned ℜ and its *témoins* in 1946. That's when he made a fateful mistake: he cleaned the kilograms *before* noting their masses and how each compared to the other. This untimely act of hygiene wiped the traceable history of measurable contamination off ℜ and its siblings. This decision of Bonhoure's to clean before measuring fundamentally changed the capacity

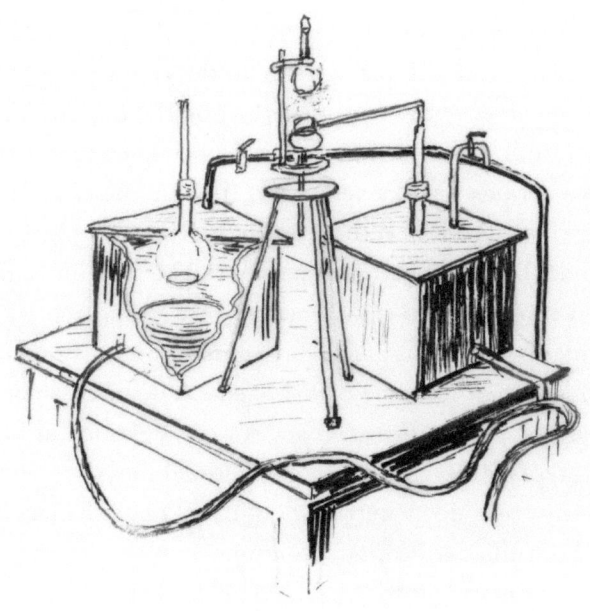

Figure 2.5

An artist's rendition of an apparatus for cleaning prototype kilograms from 1882–1889. Steam and alcohol vapor were directed alternately at the kilogram. A similar apparatus can be seen in Girard (1990), 6. Image: R. R. Mulvin.

to quantify the relationship between 𝔎 and its descendent kilograms—and forever marked the definition of mass with an uncertainty borne of bad data hygiene.[97]

Data hygiene is a set of manual protocols for maintaining the authority of data but there is also a temporality to these protocols. Once 𝔎 was cleaned, cleanliness became a part of its history—a pivotal moment that altered its commensurability with other objects. The history of its hygiene then became a crucial part of accounting for its materiality. But by mistiming the act of cleaning—by wiping and *then* measuring—Bonhoure left a gap in the documentary regime that maintained 𝔎's authority as a ground of value.

Despite this untimeliness, the BIPM used Bonhoure's new cleaning protocols on all kilograms that were subsequently sent to the BIPM for verification; the only provisional remedy to the erasure of 𝔎's history was to also scrub the objects it was compared with. There were a total of three periodic verifications of 𝔎 and the national standards that are descended from it. These periods lasted several years (the first from 1899–1911, the second from 1947–1954, and the third from 1988–1992). During these verifications, the BIPM invited members of the Metre Convention to send their national prototypes to be verified against 𝔎 and the BIPM's other working standards. By means of rotating comparison—pairing up different kilograms with each other—the BIPM scientists deduced changes of mass in national prototypes, and these changes were registered as innate errors in those prototypes. The verifications also allowed the possibility—with no absolute certainty—of deducing some estimated mass changes in 𝔎.

The third periodic verification, beginning in the late 1980s, allowed the greatest comparison among 𝔎, its *témoins*, and other nations' kilograms. It also became a staging ground for developing a more complete technique for cleaning and washing kilograms, which could be "addressed in a more searching way than had been done previously."[98] The technique developed by G. Girard, through the third verification, was interpreted earlier in this chapter in the instructions for washing your own kilogram, and describes a method for *nettoyage et lavage* that cleanses the surface of the kilogram while apparently doing no damage to the object.

In 1989, cleaning shifted from a supportive protocol in the maintenance of mass standards to being a constitutive part of the definition of mass. Henceforth, the kilogram would be defined by reference to 𝔎 *immediately after cleaning*. The full definition reads as follows:

> *The kilogram is the unit of mass; it is equal to the mass of the international prototype of the kilogram.*
>
> It follows that the mass of the international prototype of the kilogram is always 1 kilogram exactly, m(𝔎) = 1 kg. However, due to the inevitable accumulation of contaminants on surfaces, the international prototype is subject

to reversible surface contamination that approaches 1 *µg per year in mass*. For this reason, the CIPM declared that, pending further research, *the reference mass of the international prototype is that immediately after cleaning and washing by a specified method.*

The reference mass thus defined is used to calibrate national standards of platinum-iridium alloy.[99]

This redefinition of the IPK makes it clear that the mass standard is not simply defined by reference to its Enlightenment ideals—the transcendence of nature's invariants, represented by the length of a meridian—but rather by reference to the history of the maintenance of 𝔎 and the makeshift methods for mediating between its environment and its metal alloy.

Starting in 1989, the metric system was, by definition, inseparable from the protocols for maintaining 𝔎, and we can detect a historical awareness in the new definition—which is otherwise meant to be a succinct description of a basic unit—with the references to the "inevitable accumulation" of contaminants and a quantifiable amount of "reversible" contamination. This statement declares the awareness that all objects are porous and leaky: they absorb their environments and leave their own traces. In this formulation, the keepers of the metric system appear to understand and incorporate the same critique that Barad made against the whims of the scientist. The documentary history of the mass standard shows a growing awareness that 𝔎's milieu, as well as the manual protocols for its maintenance, are as much a part of the kilogram as were the finely tuned instructions for crafting the object.

Data hygiene, as an analytic term, is meant to illuminate the labor of maintaining proxies, data, and knowledge infrastructures. Initial considerations of 𝔎's material hygienics concerned its metallic alloy, the reputation of the company that crafted it, and the security of its enclosure; later in its life, having built up a history of care, new concerns developed regarding the contaminants of its environments, the traces that it picked up through circulation, and the potential to account for its history when it was compared to its fellow kilograms.

Proxies intersect and collide with their environments in unpredictable ways, despite attempts to control their circulation, to fix them as known quantities, and to protect them from interference. Before it was replaced, several people tried to deduce the source of deviations between 𝔎 and its siblings. Although it is known that 𝔎 and other, lesser reference kilograms change in mass, there is no agreed-upon explanation for their changes. One possible explanation includes the presence of atmospheric mercury used in other parts of laboratories.[100] Mercury, which may fall on floors and seep into the ground, eventually vaporizes and resettles in metals like the platinum-iridium surface of 𝔎. Cleaning cannot undo the eventual mass gains from mercury contamination. As Andrew Barry reminds us, this process is typical of metal, though counterintuitive:

> Metals are not the hard, inert objects that they are often thought to be. . . . They have become "informationally enriched," and part of the driving force for this informational enrichment comes from growing efforts to regulate the properties of the materials and the actions of those who develop and use them.[101]

𝔎 exemplifies this view—a view that we can extend to other proxies as well. The hard shell of platinum-iridium is shown to be porous, and the practices of the kilogram's handlers are deeply encoded with more than a century's worth of informationally enriched maintenance—a fact that drove the creation of a protocol for cleansing 𝔎 of this rich, lived history.

All physical proxies feature some kind of instability that produces a demand. Sometimes these physical artifacts are, by conventional decree, too instable and need to be replaced—"standards are not static, never definitions, but representations of something infinite, merely provisional drafts certain to be corrected, stand-ins for better ones to come."[102] But if they are not replaced, then physical proxies require supportive practices to keep them viable. Proxies—even those erected on a scaffolding of the French Revolution's most idealistic promises—are in constant need of maintenance. The protocols that kept 𝔎 viable lasted over many stages of its existence: its crafting, its endowment as the prototype kilogram, and its maintenance as a ground of value. Protocols for creating, storing, cleaning, and employing proxies are part of what make up the apparatus through which measurement operations become possible, and they comprise the theater of objectivity that allows a community to trust a proxy as a credible stand-in.

Whether it's the conspicuous performance of referees and officials dragging a set of chains on to the field to judge a first down in football or the trained hands of scientists rubbing a kilogram with a piece of chamois leather, ritual bolsters the significance of measurement. Cleaning the kilogram, like burying it, is not just an act meant to protect against scrutiny; it is also an internal act that signals to professionals that the work they are doing is meaningful. In commercial and regulatory standards, path dependence will often compel accordance with a standard because of the power of history, economics, and learned behavior. In the actual crafting of a standard and in its maintenance, rituals can serve a similar function by helping to conceal the fissures of arbitrariness. By developing protocols for keeping kilograms clean, officials at the BIPM invested a performance—the embodied practice of cleaning—with the endorsement of scientific necessity. These hygienic protocols were manual and practiced, and they mediated between the kilogram as a piece of scientific hardware and a piece of metal in the care of people.

Consider one of the final changes to the milieu of $\mathfrak{K}$: it moved homes. In 1889, the Prototype Kilogram and Prototype Meter were buried in a safe together, an act that compelled "cascades of rituals" that included the redefinition of the kilogram to incorporate manual cleaning.[103] From 1889 to 2002, $\mathfrak{K}$ and the witness kilograms were kept in that same safe. In 2002, the "safe was replaced by a new modern one because the old one was becoming increasingly difficult to open."[104] The problem of doors, the "hole-wall dilemma," as Bruno Latour calls it, exposes the negotiated treaties between people and technologies.[105] Such a treaty is on display in Sèvres, where maintaining a proxy as the basis of a standard was conditioned by the capacity to work a sticky door. And because the kilogram's specific milieu (including its atmospheric contaminants) will inevitably shape its physical composition, this particular door problem will have had a material effect on the makeup of the mass standard. How it changed and to what degree would always depend on where it was kept. Decisions about its milieu would not determine whether it changed, but *how* it changed. This is what institutions do: they stabilize social life and structure the behaviors they house.[106]

𝔎 was an inordinately long-lived proxy with a rarified biography. The choice of 𝔎 over other pieces of metal (or glass, or wood, or water) was arbitrary—its shape, composition, and mass were all decided with reference to a contrived set of invented units. And yet the choices that led to 𝔎, however contingent, were specially planned, designed, and treated in ways that were meant to transcend this arbitrariness. All proxies are conventional and all standards contain material aspects, but the conventional materialism that 𝔎 embodied defined the standard itself; there was no separating the metric system from 1889 to 2019 from the worldwide agreement over the material specificities of 𝔎 and the practices for keeping it as clean as possible.

As the history and end of the IPK indicate, there is a basic discomfort with the provisional nature of data hygiene protocols. They give lie to the claim that science operates hermetically and objectively by providing access "to the pure technological realm."[107] Data hygiene is a necessary, though not sufficient, condition of bringing things into measurable relationships. Although the choices of materials, shapes, sizes, and compositions may be arbitrary, those choices matter, and they solidify in forms that need maintenance. As we will see in subsequent chapters, when the choice is not *which metal* but rather *which image* to stand in for other images, or *which human* to stand in for other potential humans, the choices will have a significant impact on the composition of proxies and the potential for justice within systems of standardized knowledge.

3 THE VISUAL CULTURE OF IMAGE ENGINEERS (OR THE LENA IMAGE, PART 1)

The centerfold image of the November 1972 issue of *Playboy* magazine featured a young Swedish woman in a large, beige hat with an enormous, purple feather tassel. It appears that she is in an attic: a wicker bassinet containing a doll is visible in the background and a kerosene lamp sits over her shoulder. Her breast is exposed and reflected in a mirror on the right side of the image. The mirroring of the naked body is a generic feature of porn: it doubles the model's flesh and often reveals what is otherwise hidden from the gaze of the camera/spectator. She is naked but for the hat, a scarf, and a pair of boots—which is a weird assemblage of clothes for someone to be wearing. She stares directly into the camera. *Playboy* said her name was Lenna Sjööblom, though we now know her name to be Lena Forsén (previously Söderberg). The centerfold appeared in *Playboy* at the peak of its popularity, and the November 1972 issue was (perhaps coincidentally) the highest-selling issue of *Playboy* ever.[1]

Having considered and talked about this image for the past decade, I still think that it is a rather odd picture. The hat and the feather tassel are incongruous with the scene; the image is cropped awkwardly. The whole thing has the appearance of someone having escaped to an attic naked and put on whatever they found. That may be the point. It's indelibly marked by the aesthetics of its time, including a Vaseline-smudged lens that is unmistakably an artifact of the 1970s. It remains surprising to me that this image has circulated as an example of a *good image* for nearly fifty years–but it has.

Figure 3.1

The Lena image in its test image form. This image is called Lena_std.tif and was obtained from the Signal and Image Processing Institute's test image database. It is an excerpt of the November 1972 centerfold of *Playboy* magazine.

This centerfold would be an unremarkable footnote in the history of visual culture, photography, porn, and magazine publishing were it not for the fact that in 1973, someone at the University of Southern California (USC), in the Signal and Image Processing Institute (SIPI), scanned it to test new image compression and transmission techniques for use on ARPANET, an early computer network built by the US military and historically treated as the predecessor to the internet. Between the moment that it was digitized in 1973 and today, this particular image transformed into the "Lena image" or "Lenna image" (figure 3.1), a ubiquitous industry standard and, by anecdotal measure, the most popular digital test image of all time. The way that we look at images online was standardized by engineers and computer scientists, who often returned to the Lena image when they wanted to demonstrate new skills, new techniques, and new standards—it is woven into the fabric of our digital and visual cultures.

There are differing accounts of how a *Playboy* centerfold wound up on ARPANET in the earliest days of this network technology. The accounts agree that it took place in mid-1973 and that a SIPI engineer named Alexander Sawchuk and/or his graduate assistant, W. Scott Johnson, performed the digitization. From there, however, the precise details diverge. Version A of the origin story describes a deliberate choice to send someone out to buy a *Playboy*, which they chose because of its special qualities:

> One team member ran out to the nearest magazine store and picked up the latest *Playboy*, the fateful Lena issue. The magazine was chosen because it was one of the few publications that had full-color, high-quality glossy photos—Hugh Hefner insisted on using only the best photography and paper stock to avoid having his product considered a low-end skin rag—and its centerfold was ideal because it was the right size. Photos were wrapped around the scanner's cylindrical drum, which measured thirteen centimeters by thirteen centimeters. Folded to hide the "naughty bits," the top third of the centerfold fit perfectly.[2]

In this version, the image was specifically chosen with foresight out of a desire for a high-quality image on good paper stock—an attempt, as it is remembered, to capture the qualities of *Playboy* as a printed artifact, which were specific to its status as porn. Although someone supposedly "ran" out to a store, the scene is relatively calm and deliberate. The narrator remarks that the object and the instrument had a natural affinity—the image "fit perfectly" on the scanner—just the right size to crop out the model's breasts and scrub the image of its porniness. The image, in this version, was simply "folded" in a way that rendered it no longer illicit.

In another telling of this story, published in the *IEEE Professional Communication Society Newsletter*, the scene is much more frenzied. According to version B:

> Sawchuk estimates that it was in June or July of 1973 when he . . . along with a graduate student and the SIPI lab manager, was hurriedly searching the lab for a good image to scan for a colleague's conference paper. *They had tired of their stock of usual test images,* dull stuff dating back to television standards work in the early 1960s. They wanted something glossy to ensure good output dynamic range, and they wanted a human face. *Just then, somebody happened to walk in with a recent issue of Playboy.* The engineers *tore away* the top third of the centerfold so they could wrap it around the drum of their Muirhead wirephoto scanner, which they had outfitted with analog-to-digital converters (one each for the red, green, and blue channels) and a Hewlett Packard 2100

minicomputer. The Muirhead had a fixed resolution of 100 lines per inch and the engineers wanted a 512 x 512 image, so *they limited the scan to the top 5.12 inches of the picture, effectively cropping it at the subject's shoulders.*[3]

In this version, Sawchuk, his assistant, and the lab director used the magazine both out of necessity and an acute desire for a novel image. Instead of *folding* the image, they *tore* it. This narrative replaces the deliberate, modest folding used in the first version with a feeling of urgency. In this tale, too, there is an affinity between the scanner and the image: the engineers "wanted a 512 x 512 image," which had the effect of cropping the image above the model's breasts.

This is a fairly typical way that moments of innovation are remembered in both popular culture and in professional oral histories; moments of boredom and tedium are interrupted by frenzied improvisation. This hurried scene is characterized by the hectic pace of intense ingenuity, leaving the scanner, limited by its resolution, to do the hard work of cleaning the image of its "not safe for work" content. The irony, of course, is that by using the *Playboy* image, the engineers had already revealed that for them, the centerfold image *was* safe for their workplace: perhaps a colleague brought the *Playboy* to work, like a mystery novel, a crossword puzzle, or a newspaper to read on a lunch break.[4]

It is also apt that the Lena centerfold should follow this path to notoriety. On the formal level, *Playboy* was known as the first popular magazine with a centerfold in the United States. This is an important detail in the life story of an image often used to test the limits of data compression: the centerfold was already a compression technology that used the technique of folding to maximize the storage potential of stapled paper. The centerfold—as compression technology—employs the codec of a single gatefold to fit 50 percent more nude woman into the format of the magazine (figure 3.2). Compression allows data to travel, and the centerfold allowed the Lena image to be transported into the lab undisturbed, in the hands of an engineer.

These stories recall Eve Sedgwick's exploration of "male homosocial desire" and the (sexual and nonsexual) ways that the coherence of American,

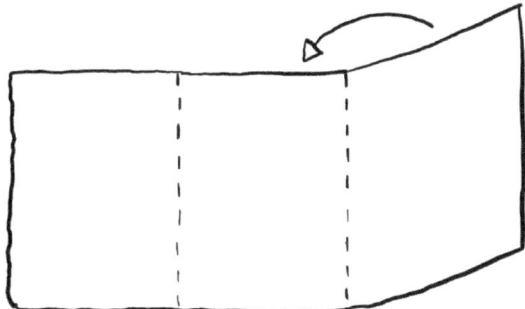

Figure 3.2

Diagram of centerfold technology. Paper folds are marked by dashed lines, with the far-right third folding inside of the middle third. The centerfold uses a gatefold to compress porn into the standard dimensions and format of a magazine. Image: Dylan Mulvin.

heteronormative desire is mediated and maintained through shared objects of desire.[5] By "homosocial desire," Sedgwick means the ways that objects like centerfolds are sites of sexual attachment—artifacts through which groups (of men) can articulate their desire and bond with other people over their shared experience of desiring. The Lena image in its original context was not just any kind of private image, but a *Playboy* centerfold, an archetypal artifact of what Lauren Berlant calls the "zone of privacy" and a conspicuous symbol of a "national heterosexuality [that] 'adult' Americans generally seek to inhabit."[6]

If the magazine was unremarkable in the USC lab, it is remarkable for this taken-for-grantedness. Even if we don't know this scene firsthand, we can recognize it as a genre of "spectacular masculinity": the garage, the closet, the shed, the "man cave"—marked by communal signifiers like the pinup calendar, the beer fridge, or a picture of a Corvette.[7] It's the way that American hetero desire left its mark on domestic and workplace architecture of the late twentieth century. And the emerging computer professions were no exception.[8] A *Playboy* centerfold in a research lab in the early 1970s was not any mere piece of pop culture detritus. It was a rank icon of normative desire and the shifting popular mores surrounding the expression of sexuality in public life.

In each version of the origin story, agency for acquiring and possessing *Playboy* was redistributed and disavowed. Desire for the image was displaced and recoded as desire for the formal qualities of the picture and the material features of the magazine—singling out the paper stock, the glossiness of the image, its dynamic range, and its portrayal of a human face. The image, which is passive in this narrative (something to simply appear; something to be retrieved), is subjected to the close analysis and gaze of the technicians. But it's the form, not the content, that they cop to wanting. This requires a significant suspension of disbelief that the content of the image (a naked woman) could be separated from its form (a glossy image on paper). It denies that the glossiness of a porn magazine is connected to the aesthetics of representing nudity. But it is altogether typical of the ways that proxies are cleaved from their origins. This is a clear attempt to clean the image, to render it as mere data and grist for a technical system.

Scholars of data are now well accustomed to noting the ways that data are never raw but always cooked, and never mere data but always material artifacts of social relations.[9] Data sets are inescapably shaped by the contexts of their collection, storage, transmission, and interpretation.[10] As part of a larger project to complicate stories about data and to rematerialize the digital, this chapter documents how an object of desire was framed as useful data *and* a useful proxy for images of the world *out there*. The Lena image served the dual purposes for engineers needing data to train algorithms and seeking a template of a human face to act as a stand-in for other faces. This story traces the social life of images in the earliest days of network technologies and illustrates how image proxies are marked by the cultural milieus of their uses.[11] By reconstructing the media practices of computer scientists in the early and adolescent periods of the internet, I excavate the norms and controversies surrounding image reproduction and the ways that test images—as porous proxies for the world—soak up the contexts of their use and reuse.

This chapter is organized around three contexts for studying proxies as porous materializations of data: the socioaesthetic context of test images used in the standardization of image technologies; the context of computer research in the 1960s and 1970s—including the regular exploitation of women's bodies; and the institutional context of military-funded image research at USC. It closes with a discussion of conversions and how we can think about the transformation of images within the larger history of proxies. Through the investigation of these contexts, I undertake an analysis that follows how the Lena image came to be and came to be taken for granted—to see how a cropped image of a white woman from a porn magazine could settle into a canonical test image for a new, digital, and networked visual culture.

It is now axiomatic in the history of media technology to recognize the ways that sexuality and technology codevelop in cycles of innovation, adoption, fear, hope, stigma, experimentation, and desire.[12] The internet is often singled out for exaggerating the effects of a "pornotroping" approach to information—one that renders and controls bodies as codified flesh.[13] This chapter and the next will contribute to that historiography. But I also want to use the institutional and technological history of image processing to understand how image engineers, in addition to being skilled laborers embedded in a university and a foundational, technical institution, were processing their labor through their identity as consumers of porn.[14] The intention, then, is to reckon with the ways that desire and control (and desire *as* control) shape technological development from the ground up.

The history of the Lena image links two forms of cultural work surrounding proxies: that of the women who have traditionally served as the models for test images, and that of the engineers, technicians, and scientists who leverage these images to build connections between their work environment, their disciplinary standpoints, and the coding of technologies. To appreciate the connection between these two forms of labor, joined as they are by a *Playboy* centerfold making its way into the SIPI labs, we first have to recognize that there is one obvious discrepancy between the two origin stories of the Lena image. In version A, a team member "ran out to the nearest magazine store" to buy the issue of *Playboy*. In version B, at the moment of technical need, "somebody happened to walk in with a recent issue of *Playboy*." But we can probably dismiss version A's timeline, as the centerfold is from the November 1972 issue, which wouldn't have been on the shelves of a nearby store or newsstand in "June or July of 1973." So maybe something like version B is somewhat closer to the truth. But this version should give us pause. Objects are never simply lying around. Porn magazines do not happen to appear in a lab at just the right moment. Timeliness is a condition of social expectations—a blend of the material culture of our surroundings and the tempos of our labor.

Objects, as Sara Ahmed writes, do not "make an appearance." Instead, arrivals take time. Objects "could even be described as the *transformation of time into form*."[15] To study the arrival of objects (or stories of their arrival) is to interrogate the contexts of their appearance, and how those contexts condition and shape what exactly arrives. We cannot understand how the Lena image came to appear on ARPANET in the early 1970s without understanding the welcome presence of the November 1972 issue of *Playboy* in the offices of SIPI. The conditions for the arrival of the Lena image had to be right. In this case, those conditions included the practiced surveillance of a woman's body by the trained eyes of engineers, who themselves were devoted to the labor of training computers in the surveillance of images. To investigate the history of a proxy test image is not only to plumb the standards of visual culture, but also to reckon with the visual culture of engineers, their position in a larger circuit of culture, and the material culture of their workplaces.[16]

The power to name—let alone *create*—proxies can shape the default conditions of a knowledge infrastructure and the common connections shared by its participants. Sawchuk, who digitized the Lena image, would go on to serve as one of SIPI's first directors, and SIPI itself would gain notoriety for its early work in image compression and analysis, as well as its database of digitized test images. If, following Marilyn Strathern, we consider culture to be "the way certain thoughts are used to think others," then the Lena image, as much as any proxy, has served this purpose for nearly fifty years—acting as a lingua franca through which image engineers could understand one another's labor and accomplishments.[17] Although the Lena image began as a stand-in for the world of images, it soon became a stand-in for the lifeworld of engineers, a stand-in for the communal media of a mostly male profession, and a stand-in for a world of images of women that could be decompiled, measured, and analyzed.

More than just an analogy, the persistence of the Lena image exposes what Charles Goodwin has termed "professional vision." For Goodwin, members of a profession shape events through the creation of objects of knowledge "that become the insignia of a profession's craft: the theories, artifacts, and bodies of expertise that distinguish it from other professions."[18] Practitioners do this, Goodwin argues, through the *coding* of phenomena, which renders everyday events into recognizable objects of knowledge in the discourse of a profession; through the *highlighting* of phenomena and their features through practices of discursive marking; and by *producing and articulating material representations*—meaning that the stuff of representation becomes "the material and cognitive infrastructure" that makes theory possible.[19] In the two narratives about the Lena image's origins, we can already see this process at work, as SIPI engineers sought to *code* a centerfold as a digital image (their domain), *highlight* its formal features (and downplay its cultural ones), and *produce and articulate* their own process of material representation—in this case, articulating the use of the image to the materiality of digital transformation.[20]

What we recognize as styles and techniques of visual representation are inseparable from the uses of image proxies, which are used to train and evaluate representational skill. From painting and drawing to three-dimensional (3D) renders and machine learning databases, shared reference images have served as tools for training and comparing the results of graphic techniques and the skills of various creators.[21] These shared reference points, in turn, serve as benchmarks for communities of practice. Chosen to stand in for the world *out there*, an image proxy becomes an artifact of a profession's history, its coherence, and a signal of insider knowledge. Through these proxies, we learn how to see, how to judge, how to classify, and how to trace our belonging in a culture.

There are at least three different, overlapping figures to consider in the Lena image—three different codings of the image. There is Lena Forsén, a Swedish person; there is Lenna Sjööblom, the name that appears in *Playboy*; and there is the Lenna/Lena image, a cropped, scanned, and digitized copy of the centerfold, whose varied spelling in the digital imaging literature indicates some of the awkward liminality of the image itself. We should be uncomfortable completely separating these three figures, as doing so threatens to undercut the humanity of the person whose body and labor are on display in *every* version of the image. However, as the history of these images and their different circulations show, in order to understand how the Lena image operates as both a proxy and a token of professional vision, it is necessary to understand how the Lena image and the model, Lenna Sjööblom, were effectively separated—how a person can be separated from her representation and how a test image is coded as a test image instead of as a centerfold.[22]

What really separates the Lena image from the centerfold is a massive act of erasure and a concerted act of data hygiene: cropped just above the model's bare breasts, the test image elides the illicit content of the original, leaving only her face and the reflection of the woman in the mirror. But the act of concealment that recodes the image and transforms it from a lurid centerfold into a decontextualized headshot can only ever be partial. The image is still marked by the soft-focus styling of 1970s magazine representations of women, and its original status as a centerfold codes the imaginary spaces beyond the cropping, as the unseen naked body haunts the excerpt. This decontextualization is also the primary act of highlighting (to use Goodwin's terminology) that transforms the Lena image into a workable proxy. The act of cropping the image above the bare breasts, whether by folding or tearing, cleansed it of its original context and transported it from a private image of desire into a testable surface.

In the context of the image's creation, we can situate clipping, cropping, tearing, and folding as ways of transforming the image and framing a vision process that would make it a viable test object. This reframing allowed the image to travel outside the immediate context of the SIPI lab. Just as the Joint Photographic Experts Group (JPEG) or MP3 standards reformat image or sound data to travel in a compressed form, centerfolds need to be reformatted as scientific objects.[23] The history of digital image compression needs to be understood through the manual labor of shaping image standards, including their test data, and the many acts of formatting and cleaning "dirty" test data for professional use.

With the growth of computer vision research driven by massive databases, there are potentially millions of test and training images, often obtained from social media and the World Wide Web. Each image has an origin story and could serve as a case study in how the material of a vernacular life can be transformed into a test object in scientific and technical research. But the Lena image is a privileged case. Not only is it excerpted from porn, and likely the most frequently used test image of its kind—with an inordinately long lifespan—it stands out from other test images because of the sentimentality espoused by engineers toward the image, as well as the woman it portrays. But even if it is a privileged case, it is far from alone. The Lena image is one in a long line of images of white women used by engineers and technicians to set the contours of so-called normality within image standardization.

TEST IMAGES: A HISTORY OF FEMINIZED WHITENESS

Now let us approach the moment of digitization in the SIPI labs as a conjuncture: a meeting of contexts that brought together a moment in American popular culture, a set of homosocial rituals, common-sense thinking, practices surrounding the use of images, and scientific knowledge-making. The first node in this conjuncture, then, is the history of test images used in the standardization of visual culture. This history demonstrates that the use of the Lena image was entirely in keeping with existing practices of calibrating visual standards to white women's skin as a prototype. And the history of test images weds the datafication of images and the visual culture of engineers, braiding together a material and aesthetic assemblage for producing standardized image technologies.

A well-used and -circulated test image is a proxy: a known quantity, a fixed point, and an invariant for testing the variables of image technologies. Test images are objects for trying things out, measuring the skills of students, and gauging the success of new techniques and technologies of image reproduction and manipulation.[24] Test images are worked upon when they are transformed, compressed, warped, masked, analyzed, identified, decompiled, and recompiled. But whatever is done to a test image must be measurable. In the parlance of digital image processing, test images are considered "data":

> Testing different methods on the same data makes it possible to compare their performance both in compression efficiency and in speed. . . . The need for standard test data has also been felt in the field of image compression, and there currently exist collections of still images commonly used by researchers and implementers in this field.[25]

In researching how scientific and technical processes choose stand-ins, the terms "test data" and "training data" frequently appear. For instance, training data can be used to hone a facial recognition algorithm's predictive assumptions by using a corpus of facial images. Test data will present the algorithm with fresh data to see if the training was successful. For this reason, test data must be data that did not appear in the training set because in order to prove that your facial recognition algorithm works on the same data set twice would be redundant and offer no predictive value of its success in the real world. The definition of "success" will be negotiated by a range of stakeholders who can contest what kinds of successes and failures are acceptable and which may be noxious.

A piece of test-proctoring software, used for administering university exams remotely, flags people of color as "unverifiable" and denies them access to their schoolwork—a situation for which students must seek redress. The technology, driven by artificial intelligence (AI), asked these students to "shine more light" on their faces.[26] A millimeter wave-scanning machine, used in airport security checkpoints and trained on a strict, binarized categorization of genders, registers statistically "anomalous" cases as suspicious, singling out some people for further inspection and scrutiny.[27] The result is greater friction, more work, more uncertainty, and less security for people who do not cleanly register in opaque, data-driven systems that are increasingly embedded within the civil infrastructures of everyday life. In each case, the likely culprit are the training and test data used to develop technologies; data that circumscribed the normative dimensions of how these technologies could be used.

In visual culture and visual technologies, using the same test images repeatedly and consistently across techniques enables a process of techno-aesthetic benchmarking, in which practitioners can weigh the costs of bandwidth or storage against the question "Does it look good enough?" or "Does it work well enough?" Benchmarking works only if we can say (in quantity or quality) how a new version compares to the original, and without a test image, difference isn't measurable. As for the Lena image's benchmarks, it features a recognizable face, a reflective surface, and a complex feather tassel that are often cited as the fixed points that computer scientists and engineers can use to track and index the success of their transformations. Test images, then, become canvases for crafting, marking, and capturing differences. As such, test images often (but not always) feature inconspicuous subject matter, clear divisions of space, and a variety of pictorial features.

My favorite test image is used in 3D object recognition, and it features a shoe, a landline telephone, and a box of miniature biscotti. It gets the job done.[28] The thinking goes that if an image technique works well on a test image or a series of test images, then it is likely to work well on future, yet-unknown images. This is possible only if a given image—like the Lena image or a shoe and a box of biscotti—is treated as a credible sample of the world of possible "natural images" (i.e., the world of rich and varied images from the vernacular world). This requires seeing the Lena image as a stand-in and imagining, however provisionally, that the way it responds to transformation and analysis will correspond with the world of as-yet-unknown images. Here, the proxy status of the Lena image is leveraged to make a wager: if a processing technique works on this woman's face, it will probably work on pictures of other faces too.

As much as we might try, it is not possible to unbraid the cultural dimensions of proxies from their material and formal dimensions. Instead, proxies must be approached as porous and leaky amalgams of their socio-material histories—they must be approached, in other words, *as culture*.[29] The uses of the Lena image, beginning with the production of a glossy centerfold for *Playboy* and continuing through its digitization and circulation among engineers, are as much a part of the image as its formal representation, distribution of features, and color values.

>>>

Femininity and whiteness haunt the history of test images. Throughout the twentieth century, images of young white women were used by engineers, technicians, and consumers to develop image standards, test that those standards were implemented correctly, and maintain their equipment. By creating technologies that more faithfully reproduce whiteness, those coded as nonwhite (especially the skin of those culturally coded as Black and Brown) are rendered less legible in image media and are subject to the compounding inequalities of intersectional prejudices, produced and exaggerated by technology.[30] In recent years, and because of the work of civil rights, antiracist, and abolitionist activists, greater attention is now paid to the failures of representation that result from using biased training data. Train your algorithms on too many images of pale-skinned people, and they will struggle to properly recognize less pale faces and flag those faces as problems for the system—a problem that is multiplied when such technologies are disproportionately used to police and incarcerate racialized populations.[31]

The failure of image technologies to render or register nonwhite skin has become a focus of activists' demands for more just image technologies. However, activists and critics have also been clear that merely using more "inclusive" training and testing data is not a sufficient response to the violences and oppressions of carceral technology. And while these concerns are magnified by the unprecedented scales of new technologies and the low level of interpretability of many algorithms, the warped representation of skin is an endemic issue in the development of both digital and analog image technologies.

As Simone Browne describes, "prototypical whiteness" operates by treating whiteness as a normative starting point and coding "darkness" as

Figure 3.3
Kodak Shirley Card (1974) portraying a woman positioned between three cushions (in the original color image, the cushions appear clockwise from right: red, yellow, and blue) and wearing a fur stole (white) with gloves (black). Original image: Kodak; photograph: From the collection of Hermann Zschiegner.

an exception.[32] The structuring power of prototypical whiteness means that some bodies are coded as legible and others as problematic. This is the ideological manner in which race becomes a "problem to be solved," to which, ironically, technology is offered as a solution.[33] Browne is building on the work of Lewis Gordon, who writes that "whites' existence is treated as self-justified whereas Blacks' existence is treated as requiring justification."[34]

Whiteness is a standard, Gordon argues, a default that can be taken for granted; adjustments to this default are exceptions and require their own explanations and justifications. But it's a process that always refers back to an inescapable, normative whiteness that structures both the standard and the exception.

Prototypical whiteness is a facet of the history of image technologies and a general "culture of light," in Richard Dyer's terms, that binds the history of pictorial representation with the history of race and colonialism.[35] As Dyer states, "white power secures its dominance by seeming not to be anything in particular"[36]—e.g., the automated proctoring software that asks you to "shine more light" on your face. The whiteness at work in test images is not an essential identity, but rather a social category that treats "white" as the unmarked and default condition of image technologies. Prototypical whiteness works in concert with other normate templates that code bodies along the axes of race, gender, sexuality, able-bodiedness, and age. Hence, whiteness is both a technological construction that encodes some bodies as more legible than others and a coordinate on a graph of social difference. In both valences it acts as a cultural adhesive, connecting aesthetic norms and technological constraints.

Because we cannot unbraid the social and the material, we have to understand the ways that they reinforce each other, in this case in the continued production and encoding of bodily difference.[37] Prototypical whiteness is baked into the history of visual media, but it extends to other imaging technologies as well, where different flesh tones and luminosities are treated unequally. Researchers have shown that self-driving cars show a "predictive inequity" in detecting pedestrians of varying skin tones;[38] that pulse oximeters—which provide vital information about a person's pulse and blood-oxygen levels—provide less accurate readings on darker skin;[39] and that fitness trackers report less accurate heart rates for people with higher levels of melanin.[40] Each of these technologies presents the possibility of worse health outcomes or death, in part because of a testing and calibration system based on prototypical whiteness.[41] By being treated as prototypical, default, invisible, and taken-for-granted whiteness escapes marking, and, in relief, defines what it means to be marked as ethnic, different, impaired, queer, or generally *Other*. In the technological construction of prototypical whiteness, otherness becomes the special case that must be explained and adjusted for, while simultaneously serving as the exception that proves the rule of whiteness's normativity. This means that "whiteness," as a cultural position and a datafied coding of flesh, is not simply captured and reproduced by image technologies; but image technologies also work within a sociomaterial system to code and crystalize nonwhiteness as difference.

>>>

Whiteness is most portable, as a benchmark for image technologies, when it is yoked with gendered representation and a normative femininity. Here the history of images of white women threads together a history of being looked at and consumed through measurement. In recent years, several researchers have taken up the history of test images to chart how engineers, scientists, technicians, and a range of standard-setters have encoded this history of raced and gendered representation within the basic infrastructures of visual culture.[42]

Genevieve Yue and Mary Ann Doane have each written about "China Girls," which, despite the orientalized name, were white women used in the calibration of film reels, from the 1920s to the 1990s. China Girls were short filmstrips, clipped from a young woman's screen test and stitched to the beginning of freshly developed film reels as tests for technicians to use in calibration. The strips would be played before the newly processed film, and through these side-by-side comparisons, technicians ensured that the film was developed correctly. In this way, they work as a fail-safe check against error. There is no clear account of the origins of the term "China Girl," and the professional use of the term apparently predates any appearance in print.[43] Despite the difficulties in precisely locating the origins of the term, the orientalist emphasis of the name privileges, as Yue writes, "a woman's subordinate, submissive behavior, qualities that would be consistent with the technological function the image serves."[44] The search for these qualities is echoed throughout the history of test images in visual culture, as well as the long-standing uses of images of women as test objects, in which whiteness and a normative femininity are used to set the conditions of new image media and maintain their default settings.

"Shirley images" or "Shirley cards" have served a similar function to China Girls, but they are used in color calibration and skin-color balance in photography.[45] Named for an early model, Shirley cards act as color bars for flesh tones. As Lorna Roth writes, skin color balance is a process in which a "woman wearing a colorful, high-contrast dress is used as a basis for measuring and calibrating the skin tones on the photograph being printed."[46] Shirley cards and China Girls are striking in the ways that they try to suture together a technical apparatus and a woman's face and body. They forecast the practices that these technologies might be used for—they imagine, however partially, the kinds of faces that might appear on film and photography stock.

Figure 3.3 (seen earlier in this section) shows a typical Kodak Shirley card from the early 1970s. It portrays a woman in upper-class garb surrounded by pillows in primary colors. Against the stark contrast of her black-and-white clothing, the image provides clearly delineated blocks of color that can be tested and compared with other images. Figure 3.4, on the other hand, is a Pixl test image produced by a Danish company, reproduced around the web (I first found it on the website of a Russian ink supplier), and used for color calibration of photo printers, acting as an unofficial Shirley card. The image features the faces of women, a pair of isolated lips, a pile of meat, a well-manicured park, a luxury car, a watch, and a naked pair of buttocks. These Pixl images circulate on message boards as calibration tools, and even though the assemblage sometimes changes (the car is updated, for instance) the women's faces and the buttocks stay constant. It's an incredible amalgam of stuff, all marked by a kind of distilled desire. Where early Shirley images dressed models in high-class finery to test image media, the Pixl image drops the pretense; women and the artifacts of conspicuous leisure all encircle a final image: a giant crevasse.

Television standards were also built on test images with similar aesthetic values as Shirley cards. For most of its history, American color television was based on something called the NTSC standard (named after the National Television System Committee). This standard was adopted by

Figure 3.4

The Pixl test image for 2009. Used for standard Red Green Blue (sRGB) calibration. Courtesy of Thomas Holm and Pixl Aps.

dozens of other countries and lasted for over five decades as a truly hegemonic standard of visual culture.[47] It's fair to describe the NTSC standard as one of the most pervasive and longest-lasting moving image standards of the twentieth century. Curiously, though, this *moving* image standard was almost completely based on still images. The engineers who built the standard used twenty-seven test images and a single filmstrip. A close look at their test images (figure 3.5) shows that they depict scenes from an idyllic, pastoral life while portraying exclusively white skin.[48] They show, among others, scenes of people boating, playing table tennis, lounging on hay, and leaning on a single-propeller airplane. Although China Girls and Shirley cards predated the NTSC, the prototypical use of whiteness as a default in image media was extended through the standardization of television.

Like the many test images that came before it, the so-called first Photoshopped image, taken and used by one of Photoshop's inventors, John Knoll, also features the half-nude body of a woman with pale skin, sitting

Figure 3.5
NTSC test image, "Boat-Ashore Pair" from Donald G. Fink and NTSC (1955), Color Television Standards: Selected Papers and Records.

on the beach, her back to the camera. The image is called "Jennifer in Paradise"; the Jennifer in question is Knoll's girlfriend, and paradise is Bora Bora. Also, like the USC engineers and the Lena image, Knoll narrates his selection of the image through the combination of its formal features, its transformability, and its affective charge. "It was a good image to do demos with," he recalls. "It was pleasing to look at and there were a whole bunch of things you could do with that image technically."[49] The image was just one of several that were used to demonstrate the capacities of Photoshop, but like other test images, it has taken on an iconic status as the ur-text of the technology.

> > >

From early film and television to the present, image technologies have been tuned to the prototype of white women's skin, used as an instrument of infrastructural calibration. This means that whiteness moves through these standards with ease, whereas darker skin creates friction for image standards. People with darker skin have historically been portrayed in less detail, with less accuracy, and in aesthetically marginalized ways. This is due to the amalgam of the ways that image technologies are tested, calibrated, and standardized, as well as the network of technologies, practices, and cultural labor that surround image reproduction. Kodak film and photography stock notoriously failed to reproduce nonwhite skin—a feature attributed to the assumed whiteness of its users *and* the film emulsion used in producing it *and* the ways that cinematographers photographed scenes *and* the ways that make-up artists were trained *and* the way that lighting professionals lit faces.[50] It's this entire circuit of people, practices, and trained know-how, calibrated through image proxies, that further entrenches whiteness as a norm of visual representation. As Dyer writes of the history of photography and cinematography,

> The assumption that the normal face is a white face runs through most published advice given on photo- and cinematography. This is carried above all by illustrations which invariably use a white face, except on those rare occasions when they are discussing the "problem" of dark-skinned people.[51]

Treating whiteness as a default meant that cases when conventional lighting techniques didn't work or some skin wouldn't register on film required exceptional solutions—these moments turned those bodies into so-called problems that exceeded the default operating conditions of the technology. Whereas Dyer documents this process in the history of film, photography, and art, we can see its traces clearly extended in the history of test images. By basing their sample of the outside world on a prototypical whiteness, these failures of imagination and consideration are embedded in technologies that are developed and calibrated according to a strictly limited cultural viewpoint.

As companies like Kodak accrued more evidence that their technologies failed to work outside of a narrow band of prototypical whiteness, they increasingly sought to remedy the problem by redesigning film standards and lighting techniques that could reproduce a wider range of skin tones.[52] These adjustments were frequently coded in racialized terms. As

Lorna Roth reports in a personal correspondence from one Kodak executive, lauding the capacities of a new film stock called Kodak Gold Max, he praised its ability to "photograph the details of a dark horse in low light." As Roth writes, "I take this to be a coded message, informing the public that this is 'the *right* film for photographing 'peoples of colour.'"[53]

If we follow Roth's reading of this correspondence as a coded message, then we can see it as an attempt to conceal the existing, racialized problems with Kodak film stock through a comparatively ridiculous scenario—there were presumably fewer complaints regarding the incapacity of Kodak film to photograph horses in low light than its incapacity to reproduce some people's faces and skin. This is not only a dehumanizing equivalence that equates (dark) horses with Black or Brown human bodies, it further denies the political demand for just representation and the dignity of having one's body faithfully captured on film.

In addition to new film stocks, film and photography companies started producing new "multiracial" Shirley cards. Figure 3.6 displays one example, called "Musicians," that was produced by Kodak and provided to consumers by an Australian photo printer. As part of a $400,000 AUD Kodak Photo CD and Pro Photo printer from the early 1990s, the "Musicians" image could be used by anyone who wanted to calibrate their own at-home monitor.[54] But on the internet "Musicians" has traveled widely and now easily can be found reformatted on message boards, hobbyist sites, and commercial printers' sites, which indicates many and makeshift ways in which proxies float through social networks, becoming recognizable through use and reuse as quasi-standard stand-ins. In this image, the whiteness and high-class garb of earlier Shirley cards are swapped out. Instead, the scene that it portrays is an emphatically ethnicized one, in which a contrast of skin tones is paired with essentializing cultural stereotypes.

Although all the models in "Musicians" hold instruments and wear a headdress of some sort, the image attempts to create another kind of striking delineation. Like the stark opposition of white, black, and primary colors in figure 3.3, here the accessorizing of the models separates them into (still-feminized) ethnic types based on an equation of skin tone with

Figure 3.6
A Kodak Shirley card called "Musicians" (1993). Image: Kodak. Courtesy of David Myers.

culturally predetermined modes of dress. For the publics of test images (photo-printers and photographers), these multiracial Shirley cards reentrench the prototypicality of Euro-American whiteness by clearly marking these remedial test images as exceptions to the rule. They situate the so-called problem of race as a problem of identity and ethnic tradition, thereby reifying the default, white Shirley as its own ethnic type.

The "solution" to the regular failures of image technologies once again materializes in the cultural, proxy labor of feminized models, here called upon to stand in for a performed diversity—a consumable otherness—that is conspicuously marked by their contrast to the unmarked whiteness of earlier Shirleys. Solutions like that presented by "Musicians" to the histories of failure within image technologies expose the limited potential of inclusion as a remedy to unjust representation—and demonstrate how inclusion itself can reinforce the representational power of the already-dominant.[55] Images like "Musicians" expose the global reach of image technologies and standards, but they do so by presenting a visualized menu of women, and evoking the commodification of otherness, in which "ethnicity becomes spice, seasoning that can liven up the dull dish that is mainstream white culture."[56]

Whiteness is a problem for technical reasons—it breaks the intended outcomes of a technology meant to portray people faithfully—but more important, it is also unjust and sets a default against which everything else must be treated as exceptional. Defaults encode the normal way that a technology is meant to work; when these defaults are tuned to normate bodily templates that are ableist, sexist, and racist, it forces the vast majority of people who don't fit those templates to adjust their behavior to fit the defaults. This is a recurrent theme in the politics of proxies. The Middletown and Decatur studies discussed in chapter 1 inserted whiteness as a default social position in measuring American experience. We see the technological construction of prototypical whiteness at work in stark relief in the history of test images, where whiteness has consistently and repeatedly mutated the conditions of possibility for new technologies and their capacities of representation.

The fact that white women's bodies—often half-clothed, naked, or in a swimsuit—are employed throughout the process of creating and maintaining a standard demonstrates how proxies are not only used as a way of modeling the world in the laboratory setting. These images suture a standard together *throughout* the process of its development, use, and ongoing maintenance. In this sense, the SIPI engineers could be confident that their use of the Lena image would cohere with the larger world of image standards, which had also been tuned to similar images of white women. Of the images discussed here, from China Girls to the "Jennifer in Paradise" image to Shirley cards, some might be used by professionals trained in computer science, image engineering, or photo processing; others might be used by amateurs and hobbyists. But the corpus braids together the sociomaterial process that forms a consistent visual culture.[57]

THE CASE OF COMPUTER VISION: THE PROFESSIONAL CONTEXT OF THE LENA IMAGE

We've seen that the Lena image—in its portrayal of a white woman's face and body—is consistent with the use of test images in other twentieth-century image technologies, from photography to film to television. On the one hand, the history of these images corroborates the notion that the Lena image did not just appear as an unprecedented kind of image in the engineering labs of USC. On the other hand, the *digitization* of a new test image was a more novel event. Although the first digital image was scanned in 1957, most test images at SIPI were reused from film and television.[58] Digital image processing was mostly new and experimental, and its practitioners hoped that it would solve many of the challenges of compression, analysis, and transmission presented by existing video and image technologies. Moreover, image technologies have often served as useful demonstration sites for the potential of new computing technologies. And this was equally the case during the period that the Lena image appeared, as many of the early and foundational experiments in AI were built with the aim of teaching a machine to recognize images.[59] As one practitioner noted in a history of this formative period in digital image work, "Almost as soon as digital computers became available, it was realized that they could be used to process and extract information from digitized images."[60]

The history of SIPI runs parallel to both the rise of computer science and the establishment of the internet—in the form of its predecessor, ARPANET. ARPANET existed from 1966–1990 as a joint scientific and military project investigating the possibilities of distributed computer networks and packet-switching. It was funded by the US Department of Defense (DoD)'s Advanced Research Projects Agency (ARPA—hence, ARPA*NET*), and led by one office in particular, the Information Processing Techniques Office (IPTO). The IPTO helped establish computer science as a discipline and provided the direction and funding of what would become the internet.[61] That background uses a lot of acronyms to say one thing: the Lena image appeared during the earliest days of networked computing, when the technology was still in flux and its uses undetermined.

During ARPANET's formative years, the research was led by Lawrence Roberts, who headed up the IPTO. Roberts, an electrical engineer, is renowned in the industry for his role in supervising the development of ARPANET and his early work on e-mail. But he had also cut his teeth as an engineer working on image processing in his graduate work at the Massachusetts Institute of Technology (MIT). The terms "Roberts gradient" and "Roberts Cross" are derived from his work and remain key concepts in computer vision and digital image processing. Strikingly, his master's thesis, "Picture Coding Using Pseudo-Random Noise," used several *Playboy* images in demonstrating a technique for sending television images over a digital network. The research also appeared as an article, including the *Playboy* images, in the *IRE Transactions of Information Theory* journal in 1962 (and is still available online).[62] It continues to be widely cited in research and patent applications up to the present day.

It was later revealed that the model in the *Playboy* images used by Roberts was sixteen when she posed for her nude photos—so she was only a child.[63] Like the Lena image, Roberts cropped the nude *Playboy* images when he transformed them into test images; but unlike those who digitized the Lena image, Roberts attributed the images to *Playboy*. Despite the revelation that his foundational study employed excerpts of underage pornography, no attempt to redact Roberts's thesis was ever made, and that fact has never been addressed or even mentioned by Roberts or by any article that cites Roberts's research (as far as I have been able to find). His research *was* anthologized by SIPI's founding director, William Pratt, who listed it in 1967 in an early bibliography of work on image compression.[64] Additionally, Lawrence Roberts's doctoral work was widely cited by the engineers at SIPI during the same years that the Lena image first appeared in their research.[65]

The appearance of *Playboy* at SIPI was not just predictable but a typical and indelible mark of the profession's values and cultural outlook. It is significant that the leader of the ARPANET project, who was responsible for funding much of the research at SIPI, was himself a trailblazing researcher in digital image technology and had used *Playboy* as his own test material. Roberts's study had already confirmed an important feature of a very new and experimental field: digital image processing would continue the pattern established in other image technologies by using images of women as test objects for the demonstration of professional skills and technical feats.

> > >

Within many of the most prominent institutions involved in the history of computing and networking, young engineers were practicing digitizing, analyzing, and transmitting images of nude women. Beyond USC and Lawrence Roberts's work at MIT, engineers at Stanford had used the communally sanctioned objectification of women and the sexualization of image analysis as a professional practice. The Stanford Artificial Intelligence Lab (SAIL) is among the earliest and most storied centers of AI research in the postwar period. Established in the early 1960s by John McCarthy, who is credited with (among many other major achievements in the field) coining the term "artificial intelligence (AI)," it pioneered research in language processing and robotics. SAIL was folded into Stanford's computer science department during the "AI Winter" of the 1980s and 1990s, a period of depressed investment and interest in AI research.

At one point in 1991, past members of SAIL circulated a remembrance of the early years of the lab, titled "TAKE ME, I'M YOURS: The Autobiography of SAIL." The remembrance is told from the perspective of a SAIL computer, a PDP-6, and focuses on notable moments in the lab's history. At one point in the message, PDP-6 discusses its sexuality (equating timesharing with promiscuity) and reminisces about a lab project in which students solicited a woman to be part of a film in which she would have sex with a computer. Having interviewed two volunteers and rejected one for being "too inhibited," they conspired to film the other volunteer sexualizing the computer while other members of the lab secretly watched on a recently installed closed-circuit television (CCTV). As PDP-6 described it:

> As you know, we timesharing computers are multi sexual—we get it on with dozens of people simultaneously. One of the more unusual interactions that I had was hatched by some students who were taking a course in abnormal psychology and needed a term project. They decided to make a film about a woman making it with a computer, so they advertised in the *Stanford Daily*

for an "uninhibited female." That was in the liberated early 70s and they got two applicants. Based on an interview, however, they decided that one of them was too inhibited.

They set up a filming session by telling the principal bureaucrat, Les Earnest, that I was going down for maintenance at midnight. As soon as he left, however, their budding starlet shed her clothes and began fondling my tape drives—as you know most filmmakers use the cliché of the rotating tape drives because they are some of my few visually moving parts.

Other students who were in on this conspiracy remained in other parts of my building, but I catered to their voyeuristic interests by turning one of my television cameras on the action so that they could see it all on their display terminals . . .

After a number of boring shots of this young lady hanging on to me while I rotated, the filmmakers set up another shot using one of my experimental fingers. It consisted of an inflatable rubber widget that had the peculiar property that it curled when it was pressurized. I leave to your imagination how this implement was used in the film. Incidentally, the students reportedly received an "A" for their work.[66]

The "experimental fingers" mentioned were likely part of two SAIL projects on robotic arms—either the Stanford Hydraulic Arm or another implement called the "Orm" (Swedish for "snake"), which "featured 28 rubber sacks sandwiched between steel plates. By inflating various combinations of sacks, the arm would move."[67] Les Earnest, the SAIL lab manager mentioned in the reminiscence, is also the likely author of this document. The image stills from this episode are still available on a website devoted to the history of SAIL.[68]

What should we make of this episode? Some might treat it as a story about students in the Bay Area in the early 1970s engaging in a sexually provocative stunt for a psychology assignment, using their research into robotics and AI as the basis of a movie not atypical of the B-movies of the era. But there is another reading of this episode that we can see beneath the glibness of the PDP e-mail. We know that the images of the woman's nude body were recorded by a PDP-10 in 4-bit grayscale and saved for (at least) the next fifty years. We know nothing about whether the woman consented to the lab's clandestine observers, the recording of her image to disk, and its storage in their database and continued storage online. As such, this episode is one of many in the burgeoning field of computer science and engineering in which the bodies of women were instrumentalized in the demonstration of technology; here, it is not the Orm or the hydraulic arm that should draw our attention, but the CCTV, networked to computer monitors throughout the lab. We know that this episode took place on March 8, 1971, shortly after the terminals were equipped with the ability to receive television signals.

Thus, this e-mail, far from a mere remembrance of the glory days of AI research, is a reminiscence about the novel CCTV system and the secret filming of a naked woman that the system enabled. These moments of self-narration tie together multiple kinds of instrumentalization—that of new technology and that of a woman's body. The episode highlights the ways that men articulate their technical achievements to homosocial desires. Like the narratives of the Lena image's origins, resourcefulness in the lab was sexualized, and the feat of putting an image of an unwitting, naked woman onto the computer network was treated as both a cultural and a technical accomplishment.

Robin Lynch has recorded other similar entries from this period. For instance, at Bell Labs in 1964, Kenneth Knowlton and Leon Harmon had a renowned dancer, Deborah Hay, pose for a nude photo, which they printed as a 12-foot-long bitmap mural that they posted on the office door of their manager. They were admonished for the prank, but much like the Lena image, they found themselves celebrated later when the image made its way into art exhibitions and reminiscences about the interconnected histories of art and computing. The image, which is known as *Nude* or *Studies in Perception I*—is often treated as "the first computer-made nude portrait." As Lynch makes clear, however, every step of the process of producing *Nude* involved a deliberate suturing together of the performance of Hay, the instruments of their computers and scanners, and the visual reference points of classical nude portraiture.[69]

The professional context for the creation of the Lena image was, therefore, the well-practiced use of nude women as test subjects and the pervasive sexualization of digital image production as an emergent technical field. Through repetition and citation, these practices of objectification fostered homosocial bonds that connected the sexualized examination of women's bodies to the professional measurement of their features. Digital image processing was a nascent discipline, but throughout many of its earliest and most prominent institutions, the instrumentalization of women's bodies was a means of demonstrating the potential of new methods of seeing.

TANK, WOMAN, TERRITORY: THE INSTITUTIONAL CONTEXT OF SIPI

The *Playboy* image that USC engineers placed on the analog-to-digital scanner was not a random excerpt from the world of pop culture. Rather, the centerfold was a conspicuous sample of the SIPI engineers' cultural milieu and a sign of the porous boundaries of the lab's environment. This milieu was characterized by the kinds of image work already happening at the USC, the precedents set by existing test images that used white women as prototypes, and an espoused desire to create a new kind of test image that reflected, through proxy logic, the cultural and technical aspirations of the SIPI engineers.

According to the institute's own description of its history, "SIPI was one of the first research organizations in the world dedicated to image processing."[70] It was established in 1971 as the Image Processing Institute (IPI) using funding provided by a contract from the DoD and the IPTO, the organization headed up by Lawrence Roberts, and the office leading the ARPANET project.[71] Prior to its foundation, many electrical engineers at USC already worked with several branches of the military, conducting image processing work contracted by the US Air Force, the Army Research Office, and the Jet Propulsion Laboratory at the National Aeronautics and Space Administration (NASA)—and indeed, much of their research concerns more efficient ways of sending images from the Moon back to Earth. As William Pratt writes, the initial work at SIPI began "on a very modest scale, but the program increased in size and scope with the attendant international interest in the field."[72]

According to SIPI's current director, Richard Leahy, "Much of the early work at SIPI was on transform coding, now the basis of the [Joint Photographic Experts Group (JPEG) and Moving Picture Experts Group (MPEG)] standards for still and video image compression and transmission over the internet."[73] The institute, in other words, aligns its early and field-defining research into image coding with some of the most pervasive and recognizable standards for digital images today. Reports from the 1960s attest to Leahy's claim. A representative report by the lab founders William Pratt and Harry Andrews describes the "classic problem" of digital image coding as "the search for a coding method which will minimize the number of code symbols required to describe an image."[74] To this end, SIPI engineers developed what they describe as a novel means of image communication: "whereby the two dimensional Fourier transform of an image is transmitted over a channel rather than the image itself."[75] When Pratt and Andrews refer to "the image itself," they are referring to a situation like NASA's Moon Surveyor missions, meant to make the transmission of images from the Moon back to Earth more efficient. Transform coding changed how this happened. Instead of sending analog television signals from the Moon, images could be sampled, turned into bits, and reconstituted from the data back on Earth.

In reports from the early 1970s, SIPI's leadership explicates the work that they are meant to be doing for the DoD's ARPA program. The reports that are contemporaneous with the Lena image refer to the institute's work as, generally, "Image Processing Research," which included every step from processing, transmitting, displaying, and analyzing to detecting and identifying images in digital form.[76] Their research goals also indicated the ways that image processing could support the aims of the DoD, if only in the abstract. In particular, image transmission from the battlefield, combined with image analysis and detection, could be used to distinguish between enemies and nonenemies. It is worth highlighting that this research was taking place in the middle of the Vietnam War and at the height of the Cold War, and that these military priorities were reflected in the test images used in SIPI reports.

Ivan Sutherland, an influential computer scientist, pioneer of computer graphics, and another early director of the IPTO, recalled the use of test images during this period in work that he called "artificial intelligence" research:

[When] I was in ARPA, the Army had *this set of tank and non-tank images,* and one of the problems was, could you recognize tanks in aerial photographs? There was this wonderful set of tank and nontank images—I think there were a hundred images. Some of the tanks were half under a tree, and some of them were recognizable mostly because of the tracks, the trail that they leave behind. For 20 or 25 years, there has been the hope that some artificial intelligence program or vision program would be able to recognize tanks reliably.[77]

I love this passage for a bunch of reasons—I especially love the idea of the exclusive categories of "tank" and "nontank" images. But Sutherland's memories of this data set are also revealing. They indicate some of the test images that SIPI engineers might be expected to use—and indeed, tank images and aerial photography reappear regularly in their research. His comments also signal the larger goals of image processing work in this period and its connection to early AI research.

Much of the early digital image processing research began as an attempt at character recognition—the idea of using a computer to process images that could be automatically "read" without help from a human.[78] The goal at SIPI was to combine this work with the ambitious project of also transmitting compressed images, potentially across large territories or even from space. Whether the "tank" and "nontank" images that Sutherland refers to are the same ones that appear in SIPI reports is unclear. But the test images used at SIPI certainly speak to the interests of the institute's government and military funders. The research was often caught between clearly military-oriented applications such as aerial surveillance and the more pedestrian uses of digital images for sending pictures of people. This awkward blending is visible in the test images from SIPI at that time. Figure 3.7 shows one example, a triptych where a white woman's face is sandwiched between an image of a tank and an aerial surveillance image.

Original Images

Figure 3.7

An artist's rendering of a triptych band of test images from SIPI: a tank, "Girl," and an aerial surveillance image of a territory surrounding water. It appears to be a port or naval base. This is an interpretation of the test image triptych found in William K. Pratt, *USCIPI Report #660*, 51. Image: R. R. Mulvin.

The image of the woman in this triptych was in wide circulation long before the Lena image and is often simply called "Girl." It is actually a frame from an earlier (1966) test film produced by the Society of Motion Picture and Television Engineers (SMPTE). Frames from the SMPTE's test film appear throughout SIPI's tests, including other images of the same woman and a "Man" image.[79] The test images at SIPI combined multiple image technologies—stitching together a new makeshift standard from the proxies of earlier technologies and the emerging applications of digital processing.

Enter, finally, the Lena image. The first time that the Lena image appears in the published research at SIPI, it is used in three different studies—suggesting its adoption as a common proxy across the lab.[80] Although it was scanned in 1973, the image did not appear in a published SIPI report until 1975–1976.[81] The first three published studies that use the image in this report are typical of SIPI research in this period, with the aim of combining the digital coding and transmission of images with the analysis and identification of their picture elements. In short, the work continues the project of combining image coding with AI and the hope that digital processing could lead to a system to automatically detect and identify picture content (like tanks!).

In addition to the sudden and widespread appearance of the Lena image in SIPI research, the lab's work alternates between a fairly stable set of test images, including the triptych of the tank, woman, and aerial surveillance images, and often running the same techniques on the triptych and the Lena images. Among the areas of research using these images were edge detection and salient feature extraction. *Edge detection* is a way of measuring discontinuities in image brightness, which allows you to outline the edges of shapes. If you were going to distinguish tanks from nontanks, edge detection would be a first step. *Salient feature extraction* is a technique for reducing the amount of information that you need to identify and describe a data set: how much of a face do you (or a computer) need to see to know that you are looking at a face and not a tank? These are both ways of looking at images and training computers to look at images and each transforms their formal components (e.g., luminosity, prominent features) into measurable data—rendering images of people and space as controllable material. The Lena image, like the images of tanks, other women, and territory, was among the material through which visual culture was becoming digitized and measurable. The desire for a "good image" meant a desire for something that could be predictably controlled.

To put a fine point on it, the test images being used at SIPI capture how the nascent field of image processing collapsed images of territories, feminized human faces, and enemy tanks into the same grammar of control. By applying the same techniques of image analysis to these three kinds of test image, side by side, techniques like edge detection and feature extraction were meant to transform images into measurable, identifiable data. Engineers were working with a stockpile of images oriented to military research goals and recycled from film and television standards. But in the genesis of the Lena image, they chose to use a new proxy for the world of images, and they chose a *Playboy* photograph as their stand-in.

Lawrence Roberts had already shown *Playboy* to be a viable source of test material, and existing test images like China Girls and Shirley cards echoed the notion that images of white women could be useful stand-ins. Choosing an image from which to extract features requires a familiarity with the features that one wishes to extract. It means being able to choose and desire an image that one can consume as an *object*. Tasked with training a computer to understand salient images, a homosocial and (almost) universally male group of engineers selected an artifact of American mainstream porn to train computers in a new science of recognition.

As historians of computing and networking have shown, much of the early history of the internet is concealed through secrecy, a lack of publicity surrounding the IPTO, and a lack of interest in its operations.[82] This lack of traces reflects my own experience with this era as well. Janet Abbate's history of the internet is a valuable resource, but it doesn't mention SIPI.[83] The piecemeal documentation of early ARPANET work makes it difficult to understand this period outside of the dominant narratives provided by its main architects in interviews and existing histories. Those narratives are characterized by the frequent claim that a series of high-risk decisions about packet switching, time-sharing, and distributed and open networks created the internet as we know it. It is difficult to test these claims with a limited set of documentary evidence from ARPA, the IPTO, and other involved institutions.

What do we make of SIPI's work on ARPANET, in the midst of the Vietnam War? Looking to the historiography of the internet shows a denial of a connection between the two. Abbate writes:

> One potential source of tension that does not seem to have arisen within the ARPANET community was the involvement of university researchers—many of them students—in a military project during the height of the Vietnam War. *It helped that the network technology was not inherently destructive and had no immediate defense application.*[84]

It is worth investigating the claim that both practitioners and historians of computing treat work on ARPANET as "not inherently destructive." First, if this were the entire case, then we might expect that work on ARPANET would be immune to the backlash against military research in universities during the Vietnam War. However, as Bob Kahn (one of IPTO's directors in this period) says, the IPTO's budget was suppressed in the mid-1970s due to the "Vietnam Syndrome,"[85] meaning *someone* (if only the DoD) thought that the work wasn't shielded from the war or blowback from the war.

Second, to argue that "the network technology was not inherently destructive" is to take a narrow view of network technology. For starters, multiple members of the IPTO research team point to the IPTO's involvement in crafting command-and-control technologies, with direct application to destructive activities.[86] This includes the image processing work going on at SIPI and the use of AI to distinguish between tanks and nontanks—with the direct implications for choosing bombing targets more efficiently.

Third, if we understand the work at SIPI as part of a larger trajectory of research into surveillance and camouflage—a historical struggle between *hiding* and *detecting*—then it is harder still to exceptionalize it as nondestructive.[87] In the Vietnam War era, image processing became the newest technique of beating the enemy's camouflage strategies; this destructive reality is materialized in the database of test images, including tank and nontank images, the Lena image, images of other women's faces, and images of territories. The fantasy of digital image processing was one that combined an entire vision process, from the moment of observation through transmission, and up to the moment of identification and analysis. The dream is that to see is to know and to know is to control. Test images outline this process by standing in for the people, places, and things that could be seen, known, and controlled.

CONVERSIONS

The history of test images is one in which the cultural labor of models, acting as stand-ins, is leveraged by engineers, scientists, and technicians, who exercise their power to delegate stand-ins for the world of images *out there*. Test images function as a stable set of pregiven data for image professionals to use to demonstrate their aptitude and skills. In addition to being a pregiven set of data, test images are the basis of commonality and community; they embody a sameness that enables the measurement of difference.

To become an industry standard and a pregiven set of data, the Lena image underwent three kinds of conversion: from paper to pixels, from analog to digital, and from the standards of a soft-core porn magazine into an image standard. When asked, SIPI engineers say that they converted the Lena image from a centerfold into a digital image as a response to their work environment and a desire for a dynamic image with distinct formal properties on "good paper stock."[88] They characterize their work environment as boring and their labor as repetitive, and they claim to be "desperate for a new test image"[89] and "tired of their stock of usual test images, dull stuff dating back to television standards work in the early 1960s."[90]

We now know that their existing images were either recycled from previous standards or dictated by their research goals. As a matter of labor, the engineers responded with sheer boredom to the repetitive use of their existing set of images and the Lena image appeared as a potential answer to boredom. A welcome break from things like tank and surveyor images, it was a chance to make a new kind of test image. The Lena image was not a haphazard fluke of engineering: it was the result of a concerted effort to crystallize the existing and familiar standards of porn's pictorial style in a new technology.

The Lena image is frequently lauded for its formal features—often in the process of disavowing desire for the woman it portrays. In the Jargon file (a glossary of computing terms first compiled and hosted on ARPANET in 1975), the image is described as having "interesting properties—its complex feathers, shadows, [and] smooth (but not flat) surfaces," and these properties are regularly cited in technical justifications for the image's digitization and its continued use today.[91] The implicit argument goes as follows: it is not that the Lena image was simply a centerfold; it was also a particularly testable image that gave engineers a set of problems to solve, including complex surfaces, reflections, and overlapping textures. This logic is repeated as new image techniques are often modeled on the Lena image, even when those techniques were developed using a much larger set of test images.[92]

But the story of the Lena image's appearance in the SIPI lab is about so much more. When someone is said to have "walked in with a recent issue of *Playboy*," what we're getting is in fact a story of the material portability of compressed data and the ways that data can travel. The history of the Lena image's conversion, then, is a series of stories about how an image is transformed and converted in order to enable its circulation and to make it portable. If no one compresses the Lena centerfold by folding it inside the magazine, it can't be carried around an office by hand; if it isn't compressed digitally, it can't be transmitted over a network; and if isn't cropped of its explicitly sexualized content, it can't be used as a scientific instrument. It is all these acts, not just one in isolation, that lifted the Lena image from the private sphere of desire into the sphere of professional image analysis. Objects have to be made portable, but some objects can be moved more easily. It's necessary to question why the Lena image moved so easily into the lab environment, through a new digital network, and into the pages of disciplinary journals. The ease with which it flowed from private domain to research network demonstrates that it already possessed some of the necessary affordances to move between environments as a ready-at-hand proxy for the world of images.

At the advent of networked computing, in a moment when engineers at USC were testing the potential of sending images as data, they chose to encode *Playboy*'s aesthetic template within their new medium. The centerfold was the most conspicuous section of an iconic magazine, chosen at a moment in which American porn was going mainstream. Not only does the Lena image appear in the highest-selling issue of *Playboy* ever, that issue appeared only a few months after the release of *Deep Throat*, one of the first hard-core porn films to be seen by a significant portion of the American public.[93] And, in 1973, the same year that USC engineers digitized the Lena image, the US Supreme Court delivered a decision in *Miller v. California* that rewrote the limits of acceptable pornography by redefining the meaning of obscenity. The court said that while obscenity had been defined as something "utterly without socially redeeming value," it would henceforth be defined as anything lacking "serious literary, artistic, political, or scientific value."[94] The ruling in *Miller* actually broadened what could be considered obscene and was widely viewed as a reaction to the recent popularization of mainstream porn. The new definition would narrow what was acceptable and widen the ambit of the state to censor sexualized texts.

But the implications of this timing are significant. In the year of the Supreme Court's ruling in *Miller,* the engineers at SIPI inadvertently helped *Playboy* overcome the new standard for obscenity by incorporating the magazine's centerfold into the knowledge infrastructure of their profession; by doing so, they relicensed the use of porn—already established by Lawrence Roberts—and demonstrated its potential scientific value by converting it to a test image. By inscribing *Playboy*'s aesthetic standard into the prototype of the internet, SIPI engineers highlighted the image as a professional object of study and encoded the practice of reading pornographic images as a professional commitment.

The history of test images from photography, film, and television through to digital image processing shows that the instrumentalization of women's bodies was part and parcel of creating a new image standard. The Lena image continues this patterned, historical recurrence and reencodes a feminized whiteness as the prototype of image technologies. In the 1970s, with an emerging prototype of what would become the internet, computer scientists and image engineers were hard at work perfecting the most efficient and dependable ways of sending a now-cleansed bit of porn over a computer network.

The origins of digital image processing were not inevitable. While the Lena image ascended as a constant against which the variabilities of different image techniques were tested, it was always superseded by a more forceful constant: a model of professional vision conditioned by the prototypical whiteness of test images and a field shaped by the pursuit of controlling space and bodies through optical capture. The researchers at SIPI often focused on new techniques for managing warfare, analyzing space, and classifying enemies. They frequently ran this line of study alongside the segmentation, identification, and analysis of women's faces and bodies. Digital image processing is a continuation of a longer history that weds militarization and the control of women's bodies, and the practices of professional vision traced by the Lena image force us to see these two forms of optical control as inseparable.[95]

To talk about the politics of the Lena image is to, by extension, talk about image standards like JPEG and MPEG that its use helped establish. The politics of test images compel us to ask, "Who is seeing?" "What is this a representation of?" "Who is it for?" and "Who gets to use it?" Image standards are shaped by testing regimes and the practices of professional vision, which are in turn shaped by the cultural milieus where professionals work. In chapter 4, as we follow how the Lena image circulated outside USC, the image's history traces the contours of a new discipline. The Lena image was exchanged, cited, lauded, canonized, challenged, resuscitated, and finally abandoned, as its circulation continued to expose who had the power to choose proxy images, how those images were used, and which contexts those images sought to represent.

4 PROXY JUSTICE (OR THE LENA IMAGE, PART 2)

An article in the 1997 issue of the *Electronic Engineering Times*, written by engineer and journalist Sunny Bains, begins:

> The most famous female face in the field of electronic imaging was honored last week. Engineers were urged to meet her and get her autograph. Conference speakers were encouraged to include her work in their publications. What did this woman do to gain such respect and admiration? Did she win a Nobel Prize? Was she voted the best Ph.D. supervisor in the country? Did she invent a device or algorithm that earned her company oodles of money?
>
> None of the above. In fact, she took off her clothes for *Playboy* a couple of decades ago, and a (presumably male) reader decided that her centerfold would make a great subject for some image-processing experiment.[1]

By the 1990s, the Lena/Lenna image was no longer a curiosity of test media from a southern California lab. It could now be called the "most famous female face in the field of electronic imaging"; however, as this passage hints, its notoriety as both an industry standard *and* a centerfold was beginning to create friction. Bains herself was critical of the image's use, and the tone of the passage indicates her disdain for the context that created it.

In chapter 3, we examined the origins of the Lena image (see figure 3.1)—a photo of a woman's face, partially covered by a hat and a feather tassel. The image was cropped from the centerfold of the November 1972 issue of *Playboy* magazine, where the full image shows "Lenna Sjööblom" naked in an attic (the spelling of Lena/Lenna has varied over time). When

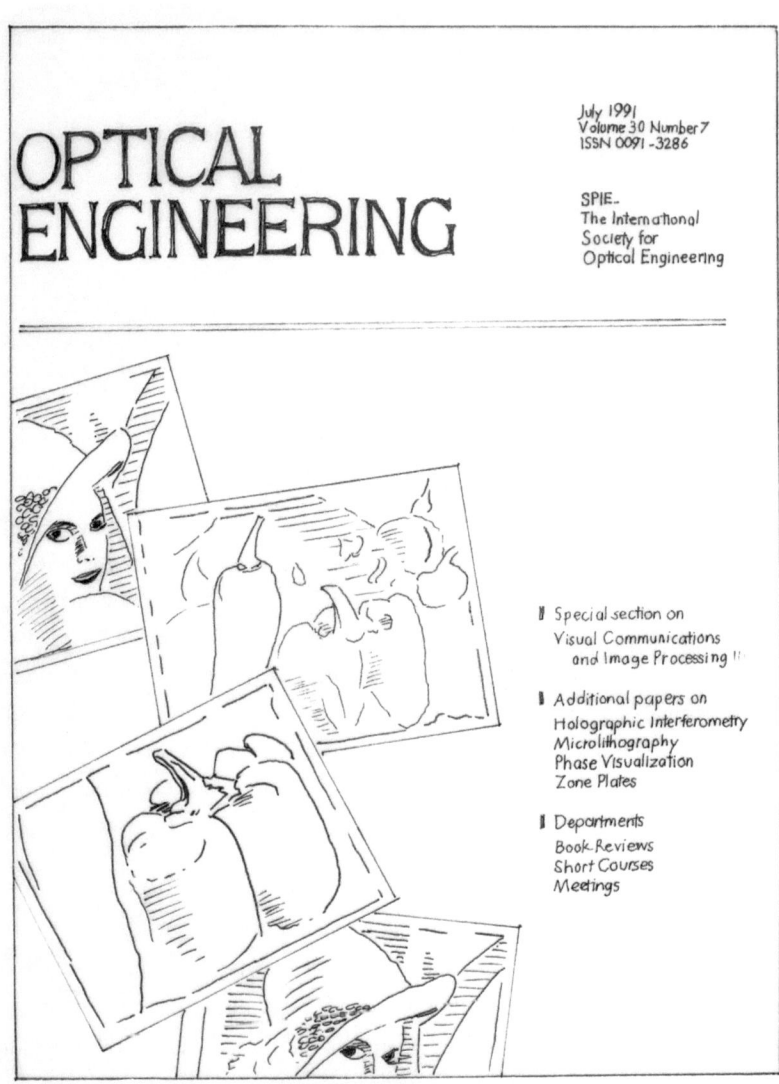

Figure 4.1

An artist's interpretation of the cover of *Optical Engineering* (July 1991), featuring the Lena image and another widely used test image of bell peppers. Image: R. R. Mulvin.

engineers at the Signal and Image Processing Institute (SIPI) at the University of Southern California (USC) digitized the image in 1973, they created a new proxy for digital image processing.

SIPI went on to shape the digital image standards of our contemporary world and the ways that images travel and appear in networked communication. All the while, this work was tested and displayed through the image of a *Playboy* model. This background to the institutional and cultural context of the Lena image, as well as its position within the history of test images, are marked by the prototypical encoding of whiteness as a default skin tone, the instrumentalization of women's bodies as measurable surfaces, and the collapse of bodies, faces, territories, and enemy Others under a regime of optical control.

In this chapter, we fast-forward to the 1990s, to a moment when the Lena image had hardened into a material proxy for a growing field and an artifact of the trained professional vision of image engineers (figure 4.1).[2] Here, we find new institutional and professional contexts, and through key moments when the Lena image was challenged, we glimpse the activist responses to the unjust settings of computer science and image engineering. Finally, we gather some of the many afterlives of the Lena image, which in recent years has transcended its status as a proxy to become an emblem for the field of image processing and a persistent target of critique.

Although the Lena image is easier to track in the 1990s than before, there are evident traces of its movement, storage, and transmission as early as the mid-1970s. By 1977, SIPI had established and distributed its first version of the USC-SIPI Image Database, a database of test images that functioned as a storehouse of potential commensurability for the growing field of digital image processing (and is still available today).[3] The USC-SIPI Image Database worked to establish common references for the field because it could be shared, reused, and cited—it provided the network of traceable fixed points, through which a growing field could make new connections. The institute often distributed batches of test images to other research centers, in person, using data stored on magnetic tape. Alexander Sawchuk (the engineer who originally digitized the centerfold and a future SIPI director) recalls: "Visitors asked us for copies, and we gave it to them so they could compare their image processing and compression algorithms with ours on the same test image."[4] In other words, the media of digital image processing began with the circulation of privately collected images that were exchanged by hand.

A similar story is told about the origins of the computerized Bulletin Board System (BBS) in 1978. The architects of BBS, Ward Christensen and Randy Suess, exchanged programs with each other by recording them on cassettes and then mailing them. When a snowstorm stranded the two men at home, as Christensen recalls, BBS was born "of the necessity of transferring files mostly between Randy and myself, at some means faster than mailing cassettes (if we'd lived less than the 30 miles apart we did, XMODEM might not have been born)."[5] These stories of manual exchange, which follow personal networks and precede digital ones, capture two common aspects of the history of standards and proxies: at key moments, the work of building and maintaining a standard is done by hand; and the tools of standards work are assembled ad hoc, but always with the aim of establishing conditions of commensurability across space and time between like-minded researchers and workers.[6]

In 1975, the same year that the Lena image first appeared in the institute's reports, SIPI engineers transmitted test images over ARPANET.[7] While we don't have evidence of which images were transmitted at this time, the Lena image was among SIPI's most frequent test objects. Therefore, the image was spreading through two networks concurrently: by hand through the social networks of engineers whose colleagues visited the lab, and through a novel computer network that would one day come to reshape the visual culture of everyday life. The first appearances of the Lena image do not announce its use by naming or labeling it in any consistent fashion. However, by the mid-1980s, the image was sometimes labeled as "Lena" or "Lenna" in articles outside of USC.[8] It was rare for the image to be consistently labeled during this period, and it often appeared with other images of women's faces, tagged as "girl," "woman in hat," or sometimes just "womanhat." It is only when the image became the source of conflict that it transformed from a taken-for-granted piece of a knowledge infrastructure into a named object with a contestable history and politics.

A few months after *Optical Engineering* included the Lena image on its cover and in two published studies within its pages, Playboy Enterprises sent a letter demanding that the journal cease its use of the Lena image on the grounds that it was the intellectual property of *Playboy* (Playboy Enterprises is the umbrella company, launched by Hugh Hefner, that publishes *Playboy* magazine). The letter reads (*sic*):

> It has come to our attention that you have used a portion of the center-fold photograph of our November 1972 PLAYBOY PLAYMATE OF THE MONTH Lenna Sjööblom, in your July 1991 issue of *Optical Engineering* magazine. . . . Playboy Enterprises, Inc., the publisher of PLAYBOY magazine, owns the copyright in and to this photograph.
>
> As fellow publishers, we're sure you understand the need for us to protect our proprietary rights. We assume you did not intentionally make unauthorized use of our material and we ask that you contact us for authorization before using any of our copyrighted material in the future.[9]

Optical Engineering is the flagship journal of SPIE, the International Society for Optics and Photonics. The issue in question, from July 1991 (figure 4.1), advertised a special section devoted to "visual communications and image processing." A few months after receiving *Playboy*'s cease-and-desist letter, the journal's editor, Brian Thompson, had to deliver a somber message about the meaning of copyright. He wrote in an editorial:

> The image in question is used a great deal by workers in image processing and is often referred to as the "Lena" image. As SPIE noted in its response to Playboy Enterprises, *"The image is widely used in the worldwide optics and electronics community. It is digitized and its common use permits comparison of different image processing techniques and algorithms coming out of different research laboratories."*[10]

As one of the few indications about the negotiations between Playboy Enterprises and *Optical Engineering*, this passage is highly suggestive. In its appeal for leniency, *Optical Engineering* made the case to *Playboy* that the use of the Lena image is not trifling. Instead, Thompson justifies the image's use by highlighting it as an object of honed, professional vision. He qualifies the use of the Lena image by noting both that engineers have transformed the image through digitization and that its popularity, its "common use," provides the conditions of commensurability for digital image processing as a field of practice. In this pithy response, Thompson neatly explicated the role of proxies in creating coherence within domains of practice, across space, and through time. The professional vision of image engineers relies on a continuity of technical experimentation that hinges on the conventional use of a circumscribed set of images.

Thompson was also speaking for a new field that was increasingly distinct from his own. *Optical Engineering* was not primarily a journal that concerned itself with digital image processing, but rather with other optical applications like holography, x-ray lithography, and three-dimensional (3D) sensing. But in the years leading up to 1991, the journal often devoted a special issue to the burgeoning field of image processing. But what Thompson either didn't know or didn't confess was that the Lena image ran *frequently* on the journal's cover and throughout the pages of the journal.[11] In fact, the presence of the Lena image on the cover was often what marked those special issues on image processing—and marked image processing as a distinct discipline, with its own test images and reference points.

Despite Thompson's equivocating, he finished the editorial by placing the onus on researchers to manage their own copyright permissions:

> With regard to the "Lena" image, we reached an understanding with Playboy and appreciate their cooperation. However, because publishers do not know whether or not material is borrowed, adapted, etc., from other sources, be advised that *it is each author's responsibility to make sure that materials in their articles are either free of copyright or that permission from the copyright holder has been obtained.*[12]

Ultimately, in Thompson's piece, individual responsibility took the place of collective responsibility to control and manage the selection of test images. Furthermore, he does not suggest that Playboy Enterprises could be wrong. It is questionable whether the use of the Lena image as a test image, without permission, is legitimate under fair use provisions in the United States: the

Lena image is cropped and does not reproduce the entire centerfold, which conforms to the protection of the use of excerpts; it was also only ever used for research and scholarly purposes, again conforming to a use that is (in theory) protected; fair use exemptions often hinge on the use of the original being transformative and in many cases, the Lena image is, strictly speaking, transformed through various signal processing techniques. On the other hand, though, in fair use cases transformation usually refers to a creative appropriation of an original text. The fact that engineers treat the Lena image as test data, not as the grounds for creative expression or critique, might make it less likely that it would pass a fair use test. Finally, a fair use exemption would require that the Lena image plays a role that *only* the Lena image could do—for instance, in this book, only showing the Lena image could serve a discussion of the Lena image. Thompson's appeal to *Playboy* made the case for its permitted use on two grounds: it was for research purposes, and the Lena image *did* serve a unique purpose due to its common, widespread, and regular use. Its uniqueness was directly tied to its cultural role as a disciplinary proxy for the world of images, as well as the ways that it was used to foster and maintain communal connections in the field.

He did not give any details about the agreement that he reached with *Playboy* in his editorial, and he stopped short of telling *Optical Engineering* authors to cease using the image. *Playboy*, for their part, later claimed that their agreement to permit use of the image grew out of their opportunism and that they worked with the Society for Imaging Science in Technology to track down Lena Forsén so the society could invite her to its fiftieth annual conference, in 1997—the event captured in the opening passage of this chapter. *Playboy*'s vice president of new media said, "We decided we should exploit this, because it is a phenomenon."[13]

Playboy Enterprise's letter to *Optical Engineering* was the first challenge to the viability of the Lena image as a professional proxy. This controversy erupted because the image had circulated unobstructed in digital image processing for nearly twenty years, throughout which time it was used by a global community of researchers. In contrast to the image's origins, when engineers lauded it for its formal properties (its glossy paper, its dynamic range) and disavowed its content (a cropped nude woman), the first impediment to its use came from *Playboy*'s assertion that the image was their private property.

A second challenge came a few years later, in 1996, and again it emerged on the editorial page of a professional journal: *IEEE Transactions on Image Processing*, where the editor responded to accusations that the Lena image's origins in a porn magazine should preclude it from further use on sexist grounds.[14] *IEEE Transactions on Image Processing* was established in 1992, when it splintered off from *IEEE Transactions on Signal Processing* because that journal's backlog of paper submissions grew too large and the digital image processing community feared "the exodus of image processing expertise from our society to other professional groups."[15] This is a key moment in that it signals an attempt to consolidate digital image processing as a discipline: the journal materialized its separation from other disciplines to keep members within its specialization. It is noteworthy, then, that at this important point in the history of the discipline, the Lena image appears 205 times in the first volume (the first four issues) of the journal (table 4.1).[16]

For context, while other common test images appear in the journal, the Lena image is by far the most used. There are, in the same volume, six reproductions of an image of Walter Cronkite and one image of Ronald Reagan. There are virtually no traces of anyone who reads as nonwhite; even by the early 1990s, digital image processing operated through a visual culture of prototypical whiteness.[17] This meant that mastery of the tools of representation and reproduction was equated with a visual culture dominated by the oversampling of whiteness in test images (discussed at length in chapter 3). Representational equity was clearly not on the minds of engineers, even while the Lena image was used to fulfill other representational desires. As Jamie Hutchinson proclaimed in 2001, "If the criterion is frequency of Lena, then the *IEEE Transactions on Image Processing* is by far the sexiest journal out there."[18] A process of proxification began at SIPI in the 1970s, when a group of men instrumentalized the detritus of their pop culture surroundings to turn the nude body of a woman into processable data and so-called extractable features. In the pages of journals like *Optical Engineering* and *Image Processing*, we see the culmination of this process.

Table 4.1

Appearances of the Lena image in *Transactions on Image Processing* (Volume 1)

IEEE Transactions on Image Processing (Volume 1)	Appearances of the Lena image
Issue 1	62
Issue 2	61
Issue 3	45
Issue 4	37
Total	205

The Lena image was now part of the canon of test images and the chosen stand-in for a new discipline.

Citation patterns expose the unspoken politics that shape a discipline's own narrative.[19] Citation sutures and divides: it builds canons, defines boundaries for insiders and outsiders, and confers legitimacy. It first confers legitimacy on the source (e.g., an image, a text, an author), but through the networks of repeated use, it confers legitimacy on those doing the citing by signaling their awareness of a community's common ties. At moments of genesis, like the first volume of *Image Processing*, citation was a way of highlighting a shared set of referential materials that demonstrated the coherence of a newly formed field. Readers of *Transactions on Image Processing* in the 1990s witnessed how different people transformed the Lena image, and technical scrutiny of the Lena test image became an occupational obligation. Being literate in image processing meant sharing a set of professional vision practices that made one conversant in the transformations of the Lena image.

In 1996, four years into the publication of *Transactions on Image Processing*, the journal's inaugural editor, David Munson, was nearing the end of his tenure. Volume 5, Issue 1, contains a letter from the editor, in which Munson reflected on the journal's first four years and its future. On the subsequent page, however, there is a second, exceptional, editorial, "A Note on Lena." Munson began:

> During my term as Editor-in-Chief, I was approached a number of times with the suggestion that the *IEEE Transactions on Image Processing* should consider banning the use of the image of Lena.

Munson established a vague controversy but made it clear that this had been an issue throughout his tenure as editor.[20] Then he continued:

> I think it is safe to assume that the Lena image became a standard in our "industry" for two reasons. First, the image contains a nice mixture of detail, flat regions, shading, and texture that do a good job of testing various image processing algorithms. It is a good test image! Second, the Lena image is a picture of an attractive woman. It is not surprising that the (mostly male) image processing research community gravitated toward an image that they found attractive.[21]

Munson, like many before and after him, was determined to fold the Lena image into the professional practices of image engineering: he highlighted the image's formal usefulness in the specialized terms of his discipline. And he also appealed to the Lena image's value as a tool for perpetuating a chain of iteration that maintained the profession's standards. But unlike Thompson with his editorial in *Optical Engineering*, he also acknowledged that the image was popular for reasons beyond its formal features and professional history—that it was an image of a desirable subject—and in doing so, he confessed to the heterosexist logic of the image's use.

The language here is telling, as he described a nearly inexorable attraction ("gravitated") of the community toward their object of desire. Unlike the accounts of the Lena image's original digitization, this passage affirmed the sexualization of the test image; but like those earlier accounts, it disclaimed the agency and responsibility that individual researchers held for choosing their proxies. It's a rare but telling moment that undermines a common refrain in image engineering, where there is often a presumed objectivity in selecting a natural image as a proxy, because one's techniques—if properly tuned—will work similarly across the world of potential images. Munson stepped out from behind a shield of presumed objectivity to excuse his colleagues' image selection on the basis of the image's "attractiveness." Here, it was the image that had power (pull) instead of the engineers making a conscious choice.

Having stipulated the scientific and affective qualities of the image, Munson recapped the controversies surrounding its continued use. He summarized the earlier conflict between *Playboy* and *Optical Engineering* but noted that it was essentially resolved. Munson concluded:

> So what is the problem? Well, quite understandably, some members of our community are unhappy with the source of the Lena image. I am sympathetic to their argument, which states that we should not use material from any publication that is seen (by some) as being degrading to women. I must tell you, though, that within any single segment of our community (e.g., men, women, feminists), there is a complete diversity of opinion on the Lena issue. You may be surprised to know that most persons who have approached me on this issue are male. On the other hand, some informal polling on my part suggests that most males are not even aware of the origin of the Lena image! I have heard feminists argue that the image should be retired. However, I just recently corresponded with a feminist who had a different point of view. She was familiar with the Lena image, but she had not imagined that there could be any controversy. When I offered an explanation of why some persons are offended by the use of the image, she responded tartly. A watered-down version of her reply is, "There isn't much of Lena showing in the Lena image. This political correctness stuff infuriates me!"[22]

Munson was perhaps receptive to the complaints of some colleagues but took great pains to make it clear that the issue was more complicated than some may have thought. The unnamed interlocutor here played the discursive role of severing the demand for different test images from one kind of feminist politics, claiming that such demands are emblematic of a lesser brand of censorious feminist.[23] Ultimately, Munson decided that the perceived ambiguity of the issue (even the feminists can't agree!) merited a recommendation that was similarly compromised:

> As Editor-in-Chief, I did not feel that this issue warranted the imposition of censorship, which, in my view, should be applied in only the most extreme circumstances. In addition, in establishing the precedent, I was not sure where this might lead. Should we ban the Cheerleader video sequence? Should we establish an oversight panel to rule on acceptable imagery? Instead, I opted to wait and see how the situation might develop. I suspected that the use of Lena would decline naturally, as diverse imagery became more widely available and as the field of image processing broadened in scope.[24]

His equivocating speaks to a fear of shared governance as a potential curb on one's academic freedom. The fear of a "precedent" that Munson cited is a fear of shifting power dynamics and it's a testament to the fear that an explicitly political standard could supersede the power of elite scientists to choose their own test media. Munson's concluding suggestions come down on the side of the complainants, if only passively:

> In cases where another image will serve your purpose equally well, why not use that other image? After all, why needlessly upset colleagues? And who knows? *We may even devise image compression schemes that work well across a broader class of images, instead of being tuned to Lena!*[25]

His final response, then, was not to censor the image but to plead that fellow researchers might be more considerate when selecting test images. It's the final sentence, however, that makes the use of the Lena image seem inescapable. There is a hint of irony, but the notion that existing image compression schemes are "tuned to Lena!" is a clear indication of the image's power as a proxy and an adhesive that binds the profession of image engineers.

And here is the problem: Munson is drawing attention to the fact that test images do not function ahistorically. Rather, they operate through chains of iteration that maintain standards and norms. Most important, they operate through the labor of the people doing the image processing—a kind of work that is already calibrated (tuned) to this one particular image. It was not just "compression schemes" that are tuned to the Lena image; the people doing the compression calibrated their vision practices to the image. The cultural work of using and reusing the Lena image as a proxy requires this attunement. This feeling was echoed in 2001 by a Carnegie Mellon University engineer, Chuck Rosenberg, who said that "many researchers know the Lena image so well that they can easily evaluate any algorithm run on her."[26] Participation in the image engineering community of the 1990s meant putting the Lena image under scrutiny as a matter of professional necessity.

These two letters from journal editors (Thompson in *Optical Engineering* and Munson in *Image Processing*) present two ways that the politics of images, computer science, and the internet were being renegotiated in the

1990s. Thompson's letter foreshadows how the management of images as private property, and porn in particular, would shape the internet as a communication medium; and Munson's letter shows how gendered mistreatment and gendered violence were the sources of long-standing conflicts within the larger communities of computer science and engineering. These letters provide us with two pathways to studying the labor and politics of visual media and proxies, first through tracing the history of porn and property, and second through the contestation of university classrooms and computer science workplaces by feminist activists. Each of these histories, which played out in the late 1980s and through the 1990s, discloses the contested uses of emerging network technologies and the gendering of computing.

PORN AND PROPERTY

Pornography is often credited as a driving force and a shadow influence over the ways that new media and technologies are used, adopted, appropriated, and controlled. As Wendy Hui Kyong Chun states, the perception of porn's primacy was crucial to the discourse surrounding the internet in the 1990s:

> In terms of technology development, sex allegedly popularizes new devices: pornography is the "killer application" that convinces consumers to invest in new hardware. New technology is a "carrier"—a new Trojan horse—for pornography; sex is "a virus that almost always infects new technology first." Sexuality is the linchpin for strategies as diverse as entrepreneurial capitalism, censorship, and surveillance.[27]

The production, circulation, and consumption of porn helped shape the ways in which both regulation and commerce took form through the internet and the legal framework for the internet's infrastructure in the United States and elsewhere.[28] This meant that internet porn received outsized attention as a phenomenon, often crystallizing sensationalized fears about the dangers of cyberspace, cybersex, communicating with unknown others over a computer network, and the threat of children being exposed to explicit images.

The largest and most conspicuous attempts at regulating the internet were often based on efforts to sanitize a new communication medium that was both newly privatized and potentially global—to make the data of the internet more hygienic. Hence, the Communications Decency Act (CDA) of 1996 was prompted by a politicized desire to redefine obscenity and to curtail access to sexual information and sexualized media.[29] Fears of the internet in this time were not only infantilizing, they were also deeply gendered. Amy Hasinoff argues:

> In the heady rhetoric of the early 1990s, the internet is naturally democratic, anticensorship, and virtually impossible to regulate . . . Metaphors positioning the early internet as a "wild west" frontier space justified the idea that it was an unsafe place that women should avoid. The key idea is that the internet could not (and should not) be governed.[30]

The CDA was opposed by civil libertarians and many of the internet's earliest and most vociferous proponents, many of whom argued that the legislation ran counter to the free-speech principles of a flat, open, and horizontally structured network. The fact that misogyny and bigotry were endemic in many of those open communities was dismissed as an unfortunate side effect. In retrospect, misogyny and hatred have thrived online and through the architecture and affordances of the web, and white supremacist and misogynist communities continue to frequently innovate new exploits of network technologies to find new audiences, expand their reach, and evade censorship.[31] Although most of the CDA was struck down by the US Supreme Court in 1997[32]—thanks in large part to the advocacy of sexual health experts —section 230 of the Act, which shields most content-hosting companies from many kinds of liability, still structures how the internet is commercialized and used today.[33]

The Lena image is woven into the internet as a key instrument in the standardization of digital images, and it is a signal example of how the internet could be used for sending and analyzing porn. But it does more than mark yet another case where porn played a vanguard role in the development of technology. As a proxy used in the preconditioning of visual standards, it connects the visual culture of engineers and their sociotechnical practices with the visual technologies that they were building. It is not a mere token of pornographic media fought over by stakeholders in technological development; it is meant to leverage its relationship to a larger cultural context, to be used in the building of a visual internet, and to be used to determine which images would look *good enough* for a potential world of users. If the Lena image at SIPI was a stand-in for a particular kind of male desire and a world of measurable images of women's bodies, by the 1990s it had become a stand-in for the heteropatriarchal relationships that structure the spaces and institutions of engineering and computer science.

> > >

Playboy Enterprises is infamous for enforcing its copyright and going to great lengths to sue perceived violators. In the 1990s, several high-profile legal decisions followed from companies hosting *Playboy* content on digital networks (both prior to the World Wide Web and afterward). These cases include *Playboy v. Frena* (1993); *Playboy Enterprises, Inc. v. Russ Hardenburgh* (1997); *Playboy Enterprises, Inc. v. Webbworld Inc* (1997); and *Playboy Enterprises, Inc. v. Netscape Communications Corp* (2004), which went all the way to the Ninth Circuit Court of Appeal. Both *Hardenburgh* and *Webbworld* are still cited by the Recording Industry Association of America on their website's section on copyright infringement with regard to digital images.[34] In both *Hardenburgh* and *Webbworld*, *Playboy* sued because online aggregators had reproduced *Playboy*'s copyrighted images in a nonpassive way (at the time, a host that acted as a conduit had a much easier defense than one that undertook any kind of curation or control over the selection of images). In *Webbworld*, the defendants argued that they were ignorant about what their users were doing and thus couldn't be held responsible for users uploading or sharing copyrighted images. The court found in favor of *Playboy*, ruling that ignorance was not an acceptable defense in this case, as *Webbworld* had targeted adult websites and concertedly built what amounted to a database of secondhand images.[35]

Hardenburgh offers a more intricate case to consider the kinds of infrastructural and cultural labor that shape technologies and standards— including legal ones—through the management of porn. In *Hardenburgh*, a popular, preweb bulletin board system called Rusty-N-Edie's (RNE) issued tokens to users who uploaded images—the tokens became a kind of voucher for downloading other users' images. Of the approximately 50,000 images on RNE, an estimated 40,000 were porn, a portion of which were proved to be Playboy Enterprise's property. The Federal Bureau of Investigation (FBI) raided RNE's servers in 1993—which were located in the Ohio home of Rusty and Edie Hardenburgh—seizing hundreds of computers. The case became a cause célèbre for the American Civil Liberties Union, but the judge in the case used *Playboy*'s display rights to find in the magazine's favor.

In building their case *Playboy* paid an employee, a woman named Anne Steinfeldt, to join Rusty-N-Edie's and spend her workdays downloading images from their servers, finding and tagging images that were potentially owned by Playboy Enterprises. The judge describes at length the manual process of finding, tagging, downloading, and examining the pictures:

> In the early 1990s, PEI [Playboy Enterprises] employee Anne Steinfeldt was given the job of scanning on-line systems to determine whether [copyright infringing] photographs were available to subscribers via their home computers. In November of 1992, Ms. Steinfeldt subscribed to Rusty-N-Edie's BBS under the pseudonym "Bob Campbell." She conducted key word searches in the files available on the BBS, and claims to have downloaded approximately 100 GIFs from the BBS which contained reproductions of PEI's photographs. She transferred these files to floppy disks, and then delivered the disks to PEI photo-librarian Timothy Hawkins. Mr. Hawkins states that he examined the files by displaying the images on his computer monitor and comparing those images with photographs from Playboy Magazine.[36]

The work that Steinfeldt performed on behalf of Playboy Enterprises, under the name "Bob Campbell," is infrastructural labor. Just as thousands of people are now employed to moderate content on commercial social media platforms like Facebook and Instagram and to check that images uploaded to these platforms meet standards of decency (including that they are not sexually explicit), *Playboy* contracted an employee to scan the databases of a bulletin board system to locate private property.[37] Steinfeldt was already performing the manual labor of image classification on a nascent platform. She went an extra step, though, by downloading images, copying them to disks, and providing them to a librarian, Timothy Hawkins. The librarian in turn performed a final act of commensuration by comparing the images side by side with *Playboy*'s archive of images.

This analysis and memory work, all performed without the help of automation, became the basis of a precedent-setting legal decision regarding the knowing curation of copyrighted porn. While legal precedents shape the contexts in which images can legally appear and circulate over the internet, it is too easy to ignore the actual labor that it takes to create a testable batch of possibly infringing images. Unfortunately, it is impossible to know if the November 1972 *Playboy* centerfold was among the images Steinfeldt found on the RNE servers (the company identified ninety-nine images and submitted ten to the court, although they are kept sealed at an Akron courthouse). Nonetheless, it is in this context of sounding out the limits of free expression and free circulation in networked communication that Playboy Enterprises became aware of the use of the Lena image by the image processing community. The very techniques that were used by users of RNE to digitize, compress, and transmit images, as well as those used by Steinfeldt on behalf of Playboy Enterprises to build its case against the bulletin board system, were built in the same laboratories that used the Lena image as the basis of technoaesthetic benchmarking.

The Lena image is remarkable because of the way that it persisted over time, well beyond its original use. But it hasn't lasted as long as it has by accident; at each phase of its existence, it needed the intervention of its users to shore up its viability. In its original digitization, the act of erasure (tearing or folding) that cropped the image of its illicit content also cleansed it of *Playboy*'s trademark, printed in the bottom-right corner ("Playboy's Playmate of the Month"). And in the negotiated agreement between Playboy Enterprises and *Optical Engineering,* the image was once again saved, its use repaired, as the intervention by Brian Thompson rescued it from illegitimacy due to its being a popular proxy. The self-policing that image engineers tacitly agreed to was meant to secure the permission of Playboy Enterprises (and not of, say, Lena Forsén herself), and that permission concerned only the status of the image as *property*.[38] We can imagine another history in which image engineers voluntarily quit using the image before or immediately after *Playboy*'s copyright claim. This was the first moment that the professional community could no longer claim ignorance of the image's origins or ownership. It would then stand as a powerful counterexample to the uncontrolled distribution of pornographic images on the internet instead of the original model of the form. Instead, the image only grew more famous, while the community of image engineers was given special dispensation from *Playboy* to keep using it.

Until recently, outright banning of the Lena image was always a last resort. The only legitimate claim against the image's unfettered use that found quarter with the engineering community was *Playboy*'s legal claim that the image was private property. Like others working at the beginning of the computer projects of the Cold War period funded by the Advanced Research Projects Agency (ARPA), the engineers at SIPI couldn't have predicted the outcome of digitizing any one particular image in the 1970s—or how the primary act of cleansing the image would code its use.[39] However, by the time the image made its way onto the cover of *Optical Engineering*, there was a strong incentive to retain the Lena image as a shared reference point and a material articulation of the professional vision of digital imaging researchers.

As Judith Butler argues, our understandings and embodiment of gender norms must be iterable to persist: "The historicity of norms (the 'chains' of iteration invoked and dissimulated in the imperative utterance) constitute the power of discourse to enact what it names."[40] But chains of iteration do not simply exist as static connections; they are animated in practices of performance, citation, and memory. The use of test images as proxies is—but need not be—a powerful means of reiterating a gendered, heterosexist, and racially twisted norm within the foundations of image technologies. Standards—as technologized norms—also work as a means of shaping the representational capacities of media infrastructures and rely on proxy samples of the world *out there* as the basis for their norming logics.[41] Because the Lena image is both a de facto standard and a tool for constructing other standards, it plays a double role in perpetuating the chains of iteration and the representational norms of visual culture. But just as a norm has to reiterate to persist, breaking the chain of iteration can strain the norm. This is the political potential of refusal and willfulness. As Sara Ahmed argues, willfulness also requires a chain of action in order to break the relations of injustice: "Willfulness becomes what travels, as a relation to others, those who come before, those who come after."[42] No one mandated the use of the Lena image, and anyone could stop using it at any time. But refusal requires a system of support and reinforcement to counter the pull of normative habit.

For instance, other than its copyright status, a person could refuse to use the Lena image for any or all of the following reasons: they think that the image is sexist; the use of the image seems overly arbitrary; the image is formally insufficient; there isn't enough information about the image's production to make it a truly useful proxy; the image is too old and its aesthetics are out of date; the image, like many before it, overemphasizes white skin; and the image is not a born-digital image. Any of these may suffice as a reason to stop using the image or even to ban its use. Instead, in a final twist to this story, one engineer, concerned that *Playboy* might one day stop allowing academic uses of its image, had his wife pose for a new version of the Lena image (figure 4.2), calling it the "iLena image." The image is licensed to Creative Commons Attribution-Share Alike.

He was neither the first nor the last image engineer to invoke his wife's image. William Pratt, the first and longtime director of SIPI, released four editions of his *Digital Image Processing* textbook over a forty-year period. Each one begins with the same dedication:

> To my wife, Shelly,
> whose image needs no enhancement[43]

This constant enfolding of male engineers' wives into the image production system underscores the constant reiteration of women's images as tools of masculine mastery. While Pratt's dedication draws a distinction between the images of women whom he works on professionally (those needing enhancement) and his wife, the iLena image exposes another flawed argument: that simply removing the private copyright status of the Lena image

Figure 4.2
The iLena image is a Creative Commons reenactment of the original Lena image. (CC BY-SA 2.5 BR) Photo by Roberto Bittencourt; the model is Ila Fox.

would eliminate any danger that it poses to the coherence of the discipline's practices of professional vision. The Lena image's liminal status as private property and communal instrument provoked the first cease-and-desist complaint from Playboy Enterprises, but prizing the image from the structures of private property by negotiating its use or reformatting it using a new stand-in model does nothing to change the contexts of its creation and continued use. As activists, workers, and students have made clear, these contexts were persistently abusive and objectifying.

RESISTANCE

There is a cost to the repetitive use of women as test objects and the regular presence of porn in computer science environments. A process of objectification that played out in test materials drew an implicit connection between the day-to-day routines of knowledge production and the many forms of abuse, violence, and mistreatment surrounding computer science and engineering in the late twentieth century. It connected the violence of campuses and workplaces to the compulsory objects of professionalization.

In the late 1980s, women scientists, computer scientists, and engineers started publishing accounts of their experiences of gendered violence, isolation, and mistreatment in their laboratories, offices, and classrooms. These accounts took the form of official and unofficial reports published by universities or circulated online, among women, and through backchannels. The objections that they voiced were tied to the same social contexts that enabled the Lena image's use: the Lena image would not have made its way onto ARPANET if the November 1972 issue of *Playboy* hadn't made its way into the USC lab, and if the USC lab had not been the kind of place in which shared consumption of porn was an unremarkable social practice. As these objections collided with the growing cultural and political power of computer science, a shift to digital and networked methods of visual representation, and a drop-off in women's enrollment in computer science programs, the Lena image received public criticism as an emblem of sexist exclusion, misrepresentation, and mistreatment.

In recent years, several popular and academic researchers have reinscribed the fundamental role that women have played in the history of computing and engineering. This work documents the concerted efforts to remove women from engineering professions at moments of increasing prestige, as well as the widespread (and often intentional) failure to account properly for the constitutive labor of women, trans and nonbinary researchers, and queer actors.[44] As Mar Hicks argues, this research can revive forgotten and erased stories that are regularly concealed by an overemphasis on heroic tales of masculine dominance of technology. Beyond the recuperation of marginalized history, work still must be done to understand "how gender is a formative category for postindustrial labor markets and how gendered analyses alter the main contentions of the historiography of computing."[45] The history of proxies and the history of digital test images are avenues for talking about the role of gendering within knowledge systems and computing professions.

Learning to look at pictures of women, teaching computers to look at pictures of women, and using pictures of women (often taken from *Playboy*), all of which show white skin, have historically been compulsory parts of learning to be an image engineer. As Munson's editorial in *Transactions* notes, this is an inescapable facet of tuning one's perceptual apparatus to the technoaesthetic benchmarks of image engineering. All computing professionals experience life through gendering systems and institutions. It's necessary, then, to understand the ways that gender—as an intersectional system of representation, identity, performance, labor, and social categorization—shapes computing sciences and the work that people are asked or permitted to do.

In the early-to-mid 1990s, enrollment in science, technology, engineering, and math (STEM) programs in Canadian and American universities was dominated by men and trending away from a period of moderately increasing gender parity. Many people in these programs often found the settings alienating, inhospitable, and abusive. Particular attention was focused on the ways that gender was expressed, policed, and weaponized. In one of the most vicious examples, a man shot and killed fourteen women—mostly engineering students—and injured another fourteen people in 1989 at the University of Montreal's École Polytechnique. The gunman's motivations were explicitly misogynist; he is quoted as saying to his victims, "You're women, you're going to be engineers. You're all a bunch of fucking feminists. I hate feminists."[46] The Montreal Massacre irrevocably marked the context for women in STEM fields in Canada and forced a confrontation with the misogynistic surroundings of engineering programs more broadly.[47]

The killings and the response to them also took place during a period of increased sensitivity to the structural inequalities that produce, reinforce, and perpetuate discrimination. The 1980s and 1990s were marked by a series of very public social justice struggles that are often erroneously lumped together under the heading of "political correctness." During this period, American and Canadian campuses were often prominent sites of confrontation between those who demanded fair treatment, affirmative action, equal pay, and reparations for historical wrongs and those who rejected these claims, resisted transformative change, and treated these demands as an attack on tradition. As Joan Wallach Scott wrote in 1992, "If there were any doubt that the production of knowledge is a political enterprise that involves a contest among conflicting interests, the raging debates of the last few years should have dispelled them."[48]

Within the larger context of campus politics and social justice campaigning, a series of reports were published in this period that documented the difficulties that women were experiencing in computer science and engineering departments. These include the Spertus report from the Massachusetts Institute of Technology (MIT 1991), the Cottrell report from the University of Vermont (1992), the Winslett report from the University of Illinois at Urbana-Champaign (1993), and a cover story in *Communications of the ACM* by Karen Frenkel (1990). There were other related reports as well, including a Harvard report on *Women in the Sciences* (1991) and a second MIT report on *Family and Work* (1990).[49] Although this was a bumper crop of reports in this area, similar texts have continued to be produced since this time. There is also a collection of gray literature from this period that includes reports that are cited but not published. Where some documents were officially released by universities, many others circulated through online message boards and early social networks. This body of literature offered evidence and support for young academics and engineers looking to address the institutionalized sexism and gendered inequalities of the scientific fields in which they worked.[50]

One of the most widely cited of these reports came from Ellen Spertus, titled "Why Are There So Few Female Computer Scientists?" At the time, she was a graduate student in electrical engineering and computer science. Spertus notes that in 1990, only 13 percent of computer science PhDs went to women, only 7.8 percent of computer science faculty were women, and only 2.7 percent of the tenured computer science faculty were women.[51] Spertus's wide-ranging report documents a variety of possible factors that could contribute to the lack of female computer scientists, including social factors like stereotyping, subtle biases, gendered language, and the tyranny of low expectations.[52] Among Spertus's contributions in this report is an exhaustive bibliography, which later circulated on message boards, was added to, and became a hyperlinked web resource.[53]

The larger group of research on women in STEM fields from this period describes university computer science departments as hostile workplaces and classrooms. The reports often emphasize the pervasive display and circulation of pornography. For instance, as Spertus states, "Some computer science graduate students and staff at Carnegie Mellon were sufficiently disturbed by the display of nude pictures as backgrounds on computer terminals that they got together and tried to change the situation by publicly appealing to the community."[54] She draws on an unpublished report written by Carnegie Mellon University (CMU) students and staff called "Dealing with Pornography in Academia: Report on a Grassroots Action." After challenging the presence of porn in computer labs and appealing for change, the students and staff of CMU were met with outrage:

> The appeal closed by making clear that they were not advocating banning such displays but were requesting that people voluntarily remove them out of sensitivity to others. Responses about the appropriateness of the displays and of the appeal were mixed and are categorized in the report. Negative reactions included the position that the writers were advocating censorship "like the Nazis or the Ayatollah Khomeini," that people should not be asked to change their behavior merely because of what others might think, and that a public appeal was inappropriate but instead should have been made by individuals to individuals. . . . In response to the criticism that individuals should complain personally, several women wrote that "[w]omen asking for changes in behavior individually are exposed to ridicule and abuse."[55]

This account highlights that the requests from women at CMU recognized both collective responsibilities (creating a workplace that is comfortable for everyone) and the importance of having the power to shape one's workspace (asking for voluntary compliance instead of a policy banning porn). In response, the requests were transformed into fuel for further abuse, ridicule, and alienation, with the moderate demands for equitable and nonviolent workplaces being equated to authoritarianism. It's a textbook example of what Sarah Banet-Weiser calls the "funhouse mirror" of gendered politics, through which social justice demands are twisted and mutated to portray men as the "real" victims of inequality.[56]

At the time, MIT had a reputation for innovation, often ascribed to its valuation of unorthodox thinkers and loose surveillance of social norms. The school remains among the most influential sites of academic research into computing and artificial intelligence (AI). Among the most often credited reasons for this reputation is Marvin Minksy, who founded the school's AI Lab (now the Computer Science and Artificial Intelligence Laboratory [CSAIL]) with John McCarthy and was a member of the Media Lab from its founding in 1985. As Meredith Broussard says, "Look behind the scenes at the creation of virtually any high-profile tech project between 1945 and 2016, and you'll find Minsky (or his work) somewhere in the cast of characters."[57] As famous as MIT was for its research, it equally prided itself on its self-perceived iconoclasm. As Broussard writes, Minksy's lab was "where hackers were born. It was terribly informal."[58] It was vital to the reputation of many researchers within the AI Lab, and the Media Lab after it, that their work appear unconventional—a countercultural and counterinstitutional ethos that bolstered the claims of revolutionary potential connected with innovations in computing.[59] In recent years, both CSAIL and the Media Lab have been tarnished by their long-standing association with the convicted sex offender Jeffrey Epstein, who provided both institutional funding and personal funding and organized fundraising for institutional research. Minsky also accepted funding from Epstein and organized academic symposia on his private island after his conviction in 2008.[60]

But these institutions were already accustomed to controversy. Even in the 1980s and 1990s, MIT, CSAIL, and the Media Lab had been cited for the mistreatment of women. In Karen Frenkel's 1990 cover article for *Communications of the ACM*, titled "Women and Computing," she included anonymous quotes gathered from women computer scientists around the United States that detail experiences of harassment, abuse, and alienation. In one passage, a woman detailed her experience of choosing a proxy, a video test sequence, at MIT's Media Lab:

"[In] The Garden [at MIT's Media Lab] . . . some faculty, students, and staff [chose] a test sequence from the film clip of the TV program 'Moonlighting'. They were looking for a sequence of a few frames that had a variety of colors, textures, and camera motions, and that probably had human figures on it. On these strictly technical considerations, they chose a sequence in which, at the beginning, the camera focuses closely on the legs of Cybil Shepherd as she walks away from the camera in a torn skirt. Subsequent frames show her walking flirtatiously past Bruce Willis, pretending to be angry at him but with a small, triumphant smile on her face. . . . Women must deal with these pictures of women as test objects, as pictures to be used over and over again, long after their anger has worn off."[61]

It is remarkable how well this person's description foreshadows Munson's description of the Lena image from six years later—the way that it speaks to the repetitive injuries of using compulsory test images and the way that it captures the effects of one's embodied labor being tuned to an unwanted proxy. As much as it foreshadows Munson's comments, however, it is the twisted mirror image of the origins of the Lena image; while the man who digitized the Lena image, Sawchuk, claimed that the selection of the centerfold grew out of boredom with his existing batch of test images, this woman at MIT must deal with the tedious repetition and the perennial objectification of women's bodies. As image proxies have to be seen as typical representations of the world of images, the selection of an image that carries the signifiers of sexualized and gendered violence echoes the banality of such texts in the larger domain of popular cinema and television—as a proxy, it doesn't just index the world of pop culture media, but also the taken-for-grantedness of images of abuse.

The correspondent's description of her work in the Media Lab expresses the anger of being forced to select a sexist video clip. But it also describes the routinized violence of using the clip repeatedly as a necessary and assumed practice on the job. Such accounts are important testimonies that expose the nonspectacular ways that shared injustice must be mediated. Here too, the politics of standing-in are activated, as witnesses can speak to and speak for absent others. "Witnessing," following Carrie Rentschler, is a form of participation in others' suffering.[62] In "The Aptness of Anger," Amia Srinivasan describes "affective injustices" as a special kind of injury "where victims of oppression must choose between getting aptly angry and acting prudentially."[63] These affective injustices, where one must curb one's anger out of a fear of appearing too "emotional," constitute a form of doubled, unrecognized harm. The testimony of the researcher at MIT, like those of other contemporaneous accounts, speaks directly to the felt experience of affective injustice: the injustice of lacking control over one's working environment or the knowledge infrastructures of one's discipline, *and* the felt incapacity to do anything about it.

The concluding statement, that this labor will continue "long after their anger has worn off," demonstrates the untimeliness of unrecognized injustice: if your work has always been tuned to the stuff of prototypical whiteness and sexualized objectification, when would it be timely to be angry? Affective injustice is not evenly felt; it is more likely to be felt by those whose anger is seen as unreasonable or inapt. Affective injustices are more likely to be treated as legitimate complaints if they are voiced by people whose subjectivity is associated with rationality, not by "the sort of person who is not already stereotyped as rageful, violent, or shrill."[64] In the context of the campus politics of the 1980s and 1990s and the reactionary politics of those in positions of power, who often refused even minor demands for change, we can see how the labor of working with unjust proxies for the world *out there* persists long after the moment of acute anger.

Chapter 3 documented the institutional, cultural, and technical context of the digitization of the Lena image, and the role that human actors like Lena Forsén play in embodying proxiness; meanwhile, the discussion here has marked another crucial period in the image's history by placing it within the larger context of computer science and engineering in the 1990s. This period was marked by its attention to the structures of inequality that perpetuate discrimination and abuse within institutions, and also by the resistance to these structures undertaken by activists, students, teachers, and those simply fed up with the status quo.

Accounts like this one from MIT speak to the embodied labor of working with proxies within a larger institutional context of mistreatment and misrecognition. How is a researcher, a student, or a fellow teacher to understand their role in the workplace in such moments? It is also worth remembering that when controversies and scandals erupt, there is often an attempt to claim that things used to be different, the culture of the time was less enlightened, and things have changed now. This is usually an attempt to disavow the political demands of minoritized and marginalized people. But these scandals erupt because of the bonds of culture, the chains of iteration and reiteration that perpetuate norms. It is true (but not an excuse) that the problem is cultural—that certain thoughts are used to think others—but it is the power to create those connections, to insist on particular forms of culture that is so firmly entrenched. It's the inescapability of the network of cultural connections that makes it nearly impossible for the powerful to imagine a different way of doing things.

As Kenneth Burke writes, in an appropriately visual metaphor, "A way of seeing is also a way of not seeing—a focus upon object A involves a neglect of object B."[65] For him, ways of seeing and not seeing are learned, trained, and practiced. They come to seem natural, even valorized, as professional ways of enframing the world. Building on a term first coined by Thorstein Veblen, Burke calls these ways of seeing/not seeing, "trained incapacities."[66] He adds that "one adopts measures in keeping with his past training—and the very soundness of this training may lead him to adopt the wrong measures."[67] In other words, trained incapacity is not an unwanted side effect of proper training; rather, it is a way of understanding properly trained actions from a different perspective. This is precisely the problem that David Munson diagnosed in 1996 when he wrote, "We may even devise image compression schemes that work well across a broader class of images, instead of being tuned to Lena!" The Lena image is an instrument that tuned the professional vision of image researchers to a circumscribed world of digital images, and it also changed the ways that engineers experienced their profession.

The other story here is one of historical exclusion, a story where women had little capacity to shape computer science as a discipline, were limited in shaping the media environment of their workplaces and classrooms, and were excluded when men repaired the instrumental value of the Lena centerfold by rejecting claims of its unjust use. The Lena image was what Luce Irigaray calls the "*target, object*, and *stake*" of a masculinist discourse.[68]

The routines of professionalization in digital image processing required workers to choose images as proxies, to potentially favor certain kinds of images, and to reproduce those images with as much regularity as possible—building up the known catalog of possible image transformations. Through this process of constant reproduction, the image hardened as both a shared reference point and a useful set of test data. Understanding this repetitive labor as a manifestation of trained incapacity is one way of accounting for the selection and maintenance of the Lena image as a stand-in for the objectification and sexualization that suffused lab environments.

AFTERLIVES

Something changed in 2018, when scientific and technical journals finally started to ban the use of the Lena/Lenna image. We return to the pages of *Optical Engineering*, where the journal's editor stated, "As of 1 July 2019, SPIE journals and books will no longer consider new submissions containing the Lena image without convincing scientific justification for its use."[69] In *Nature Nanotechnology*, a similar note said, "We no longer consider submissions containing the Lena (sometimes 'Lenna') image."[70] The Optical Society and the Society for Industrial and Applied Mathematics also banned the image's use. Each of these statements claimed that this was a collective and deliberative decision and noted that the image had served as a standard for forty-five years. They also tied their decisions to a desire to create more hospitable environments for women to become computer scientists and engineers. Like other proxies, which gained coherence and usefulness through reuse, the Lena image persisted for decades as a shared reference point. But ultimately, it was political demands, not formal ones, that sunk it. Beyond just a winking reference to insiders, the image accrued a wider reputation as an icon of misogyny and misrepresentation within the world of computer science and its allied fields. What began as a campaign to end the use of the image in the 1990s culminated in a documentary, *Losing Lena* (2019), featuring Lena Forsén herself and supported by the "Code Like a Girl" campaign.

The image lives on, however. In the HBO comedy series *Silicon Valley*, the protagonists, a start-up with an enviable compression algorithm, compete against their main antagonists, a powerful Silicon Valley firm called Hooli. Hooli has both bottomless funds and an abiding grudge against the start-up. The two sides meet in the finals of a Valley competition, pitting their compression algorithms against one another. Hooli's chief executive

Figure 4.3

In this scene from the HBO series *Silicon Valley* (2014), the fictional Hooli CEO demonstrates how the company's compression algorithm would take the original Lena image (top panel) and convert it into more compressed data (bottom panel). Photos: Dylan Mulvin.

officer (CEO) stands in front of the crowd, and in the familiar bravado of the Valley's product showcases, proceeds to compress an image more efficiently than ever before (figure 4.3). That image is the Lena image.

Earlier in the series, the start-up's leader sits at his desk, trying to think his way out of a stymying compression problem. Deep in thought, his head is framed facing a now-familiar sight: the Lena image is pinned to the wall next to his desk. The image is there when the young, aspiring engineer needs inspiration, and it is there when the behemoth corporation needs to assert its unrivaled mastery: in the visual culture of Silicon Valley (and *Silicon Valley*), the Lena image is both muse and model.

Its status as an icon of the image processing profession has prompted a range of responses. In addition to appearing on *Silicon Valley* and in poetry,[71] the Lena image appears in art works, reenactments, 3D renderings, icons in Apple's App Store, and video art pieces.[72] This is not surprising. As recurs frequently with proxies, people will often find and use these common reference points as the basis of artistic creation (see chapter 6). A video artwork by Jamie Allen with the unfortunate name of *Killing Lena* shows the image undergoing repeated compression. Like a photocopy of a photocopy, the image slowly disappears in a cloud of pixelated noise while Roberta Flack's "The First Time I Ever Saw Your Face" plays.[73]

The history of the Lena image shows how the process of proxification—the transformation of a centerfold into a test image—is porous, as the test image carries the traces of its institutional and cultural milieu. It also shows how these traces leave indelible marks. Throughout the 1990s, the image's origins and continued use were liabilities that brought it into focus as an artifact of the field's larger structures of discrimination and objectification. The work of animating proxies—from their production to their continued use—is inescapably embodied. Test images capture the cultural work of models who perform as stand-ins. In being coopted into the history of test images, Lena Forsén's labor became the labor of vicarity—standing in for a world of images, faces, and skin. But the cultural work surrounding proxies also extends to the moment that this labor is leveraged, like the moment when image engineers walked into the SIPI laboratory and either tore or folded a *Playboy* centerfold to fit it with a scanner; the conditions of proxy labor also extend to the malaise of using a staid set of test images, as well as to the alienation of compulsory engagement with sexist and objectifying instruments. Proxies live through the bodies of their users.

Proxies for the world *out there* don't just form the foundation of a discipline, but also themselves are in need of maintenance and repair and are susceptible to contestation. For it to stay usable, someone had to crop the Lena centerfold to cleanse it of its soft-core origins, someone else had to circulate the image to turn it into a commonplace object, and someone (indeed, many someones) had to use and cite it. In the making of standards, cropping, circulation, and citation all become kinds of repair and maintenance. These social acts shape and maintain proxies in ways that lend them credibility and allow them to retain legitimacy.

Our media technologies are built on standards, and our standards in turn are built of the materials that standard makers use. The history of proxies like the Lena image, however, tells us that our standards are made to work in the world, but they are made to work for only some people in some ways. Objects do not simply appear and centerfolds do not simply appear on analog-to-digital scanners. To appear, an object needs time, circumstance, and purpose.[74] Test images appear because their users want to forecast what they think their publics and their machines will see. They imagine *how* we will see—whether it's the automated detection of tanks, nontanks, or centerfolds, test images foresee one version of things to come. We know that our ways of seeing are learned, that human vision is trained.[75] We also know that computer vision is trained. The Lena image, as an early digital test image and a product of the material life of hetero desire, serves as a familiar icon of both these processes.

There is a difference between what we know and what we see. The myth of computer vision and digital image processing is that they are the same thing—that a computer or an algorithm might "see" in an objective way, unalloyed by human perception and prejudice. Yet these techniques function through specifically chosen and programmed preconditions. The history of the Lena image tells a story about humans caught up in data. It's an entangled tale of gender, sexuality, race, and power: the power to choose test media, to inscribe new techniques of vision, and to dictate a new vocabulary of seeing. But it also tells a history of resistance to that structure of vision, and it demonstrates that the meaning of a standard is never fully determined.

5 LIVING PROXIES: THE STANDARDIZED PATIENT PROGRAM

To train medical professionals in proper care, the standardized patient program uses actors to simulate illness and disability. Once given an assignment, the "patients" are interviewed and examined by medical trainees seeking to be nurses or physicians, who try to diagnose the performers. The entire process is monitored, and the trainees are evaluated on their methods, the accuracy of their diagnoses, and the compassion that they demonstrate during their interactions with the standardized patients (SPs).

In a radio profile of an SP based in Los Angeles, the host captures the program through an analogy: "Like the nude model in art class, or the customer at the beauty school salon, [the SP] is a human *learning tool*."[1] Similarly, Janelle Taylor quotes a typical speaker at an SP conference: "this is a *technology* from my perspective. Whether you're using a person, or a computer program or a mannequin, it's all part of the continuum of simulation."[2] Living human bodies, behaviors, and biographies complicate these analogies between life models, mannequins, tools, and technologies. This is because—to dispute the previous statements—people will always resist attempts to make them into simple tools and technologies. SPs offer their bodies as stand-ins for a world of potential bodies, performing their humanity while simultaneously acting *as if* they are sick or disabled. But their bodies linger, and the standardized patient program must creatively manage the overlapping humanities of those who act as proxies and the people for whom they stand in.

The standardized patient program is part of a shift in medical training toward the clinical performance of care and the standardization of the "emotional labor" of medical professionals.[3] Today, every physician in the United States and Canada, whether trained nationally or internationally, must complete an exam called an Objective Structured Clinical Examination (OSCE) based on SP interactions.[4] Medical licensing and the possibility of standardized medical care now hinge on a bureaucratically administered system for training actors to embody scripted roles, simulate disease, and perform as disabled.

Proxies are relational artifacts, meant to bind institutions and disciplines to shared reference points. They animate the work of culture that uses things—or people—to think other things. We've seen how proxies, from the most basic pieces of metal in the metric system to the stuff of a male-dominated image lab, are porous, leaky, and unstable. All of these issues threaten the long-term viability of proxies and, with them, the knowledge infrastructures of these institutions. The history of proxies is as much that of their selection as of their maintenance, upkeep, and repair. The history of the standardized patient program, accordingly, is a particularly thorny one.

This story of proxification traces the way that human bodies—in all their leakiness and porousness—have been trained by a medical establishment to act as relational instruments. SPs are meant to stand in for the experience of illness and disability in order to encode a form of medical care that lives up to the standards and expectations of the profession. The program began as a way of imagining a better relationship between physician and patient, and it continues as a way of simulating those relationships in the reproduction of medical and caregiving norms. By simulating doctor-patient interactions, the standardized patient program is meant to train people in the embodiment of illness in order to compel physicians to embody *proper doctoring*. This relational dynamic, for which there could be no computerized or written substitute, makes human proxies invaluable in the standardization of care.

This chapter has a number of goals. First, it conveys a brief history of the standardized patient program as it took root in the United States and Canada—its transition, over a few decades, from a curiosity into an entrenched aspect of medical education and accreditation. Second, it paints the standardized patient program as a bureaucracy that relies on human beings who can act as faithful, consistent, and trustworthy proxies for the pain and suffering of others. Third, and finally, it shows how all the cultural practices surrounding proxies, including the use of fixed points, the ritualized performance of embodied gestures, the vesting of particular materials as stand-ins, and the inescapable regimes of data hygiene, apply when the target of standardization shifts to human bodies and human interactions. In addition, this chapter charts a novel process of measuring, quantifying, and standardizing affect and care in the late twentieth century. Such efforts take place within a larger context of the quantification of emotion, affect, and care, as well as the development of new, intricate relationships between measurements and feelings.[5]

A NEW STANDARD

Between the 1970s and 2000s, medical educators in Canada, the United States, and Great Britain adopted a greater performance-based framework in training and accrediting physicians, and the standardized patient program is just one example of this process. This period was also characterized by the influence of medical educators trained in psychometrics, which has led to more emphasis on "standardization, reliability, and validity in assessment."[6] This shift in emphasis was also compelled by the rise of a patient safety movement, an increase in malpractice lawsuits filed against American physicians, a perceived increase in medical errors, and a decrease in patient satisfaction with care.[7] A shift to a performance-based framework demanded a new standard—and corresponding simulations—that could be used to evaluate any physician who wanted to work in the United States; today, OSCE testing with SPs is currently centralized in five US cities, and exams are given every workday.[8] This new licensing requirement emerged from a belief that it was financially and ethically necessary to supervise the diagnostic skills and affective care of physicians.

The use of SPs dates to 1963, when a physician, Howard Barrows, and a medical education researcher, Stephen Abrahamson, at the University of Southern California (USC)—the same campus where, exactly a decade later, the Lena image would first be digitized—were looking for a way to train and compare third-year medical students' interactions with patients. Barrows and Abrahamson were working in a context where little attention was paid to the training of "bedside manner" and patient compassion. A contemporary test of bedside manner was scrapped by the National Board of Medical Examiners in 1964, when it was revealed that the evaluation results were as random as chance.[9] Early adopters of Barrows's standardized patient program, who used it as a supplement to their gross anatomy classes, claimed to view the SPs as "bridging the gap between cadaver anatomy and the anatomy of the living human being."[10]

Barrows was a young physician in USC's Department of Neurology, and his collaborator, Abrahamson, was the school's director of the Division of Research in Medical Education. They published their first paper on the use of SPs in 1964 under the title "The Programmed Patient: A Technique for Appraising Student Performance in Clinical Neurology."[11] Over the years, the techniques that Barrows and Abrahamson developed evolved from using the term "programmed patients" to "simulated patients" to, finally, "standardized patients."[12] The change in name does not indicate a drastic change in either methodology or the demands placed on the actors involved. However, the choice of which of these words to use— "programmed," "simulated," or "standardized"—says something important about proxies: they are *programmed* objects meant to *simulate* an event for the purpose of *standardization*. By ultimately calling the proxies SPs, however, the process is actually reversed: they are *standardized* instruments meant to *program* medical students in proper behavior through *simulation*.[13] The Canadian psychometrician Geoffrey Norman, who coined the term "standardized patient," did so for this exact reason: he wanted a term that captured the "technique's strongest features, the fact that the patient challenge to each student remains the same."[14]

THE SCENARIO

Standardized patient programs use a preplanned set of scenarios, within which SPs embody fictional patients with symptoms, past behaviors, and a set of questions to ask their trainee physician. There are several variations on what a trainee may be asked to do, but in general, they will enter the scenario *as if* it were a normal clinical exam. The trainee may be asked to take the patient's history, perform a physical exam, and try to form a working diagnosis. Throughout the scenario, they are meant to gather information while expressing proper compassion toward the SP. Study guides for physician trainees universally emphasize the importance of being polite, gentle, and presenting a neat and clean appearance. Almost every guide begins with a reminder to trainees not to forget to wash their hands.

Trainee physicians encounter SPs in a circumscribed scenario: they enter a room and are either alone with the SP, watched on a closed-circuit television (CCTV) system, or perform in front of colleagues and teachers. They encounter the bodies of the SPs via their sight, hearing, and touch and all of these senses help build a network of inferences, analogies, experiences, and likenesses. In *The Birth of the Clinic*, Michel Foucault describes the logic of this kind of encounter as a product of the eighteenth-century emergence of a "clinical" or "medical gaze," and with it a shift to the primacy of the physician as the locus of knowledge about disease. The clinical gaze centers the physician's understanding of symptoms as a "combinative study of elements," by which a series of symptoms would add up to a diagnosis of a distinct disease.[15] A principle of analogy (the conditional *as if* of a patient's history) displaced the patient's own perspective and recodes the body as a constellation of information in a series of patterns. As Hsuan Hsu and Martha Lincoln describe: "Facilitated by the medical technologies that frame and focus the physicians' optical grasp of the patient, the medical gaze abstracts the suffering person from her sociological context and reframes her as a 'case' or a 'condition.'"[16] The standardized patient program could not operate without the deep, institutional, and infrastructural normalization of the clinical gaze. As a simulation meant to train the physician in apprehending a probable diagnosis—*whether or not the disease actually exists in the person*—it operates on the logic of a case study, where patient descriptions situate their bodies on a graph of similarity to other bodies.[17]

To generate the SP scenario, medical educators first develop a profile of a case study using a template and then distribute patient scripts to the

Patient Behavior:

Affect: (check all that apply)

[ ] relaxed	[ ] cooperative	[ ] pleasant	[ ] confident
[ ] uncooperative	[ ] hostile	[ ] demanding	[ ] preoccupied
[ ] anxious	[ ] fearful	[ ] apprehensive	[ ] sad
[ ] listless	[ ] sad	[ ] withdrawn	[ ] other_____

Body Language:

| [ ] relaxed | [ ] withdrawn | [ ] defensive | [ ] uncomfortable |
| [ ] anxious | [ ] fearful | [ ] nervous | [ ] other_____ |

Facial Expression:

| [ ] relaxed | [ ] tense | [ ] worried | [ ] irritated |
| [ ] other_____ | | | |

Eye Contact:

| [ ] normal eye contact | [ ] looks away frequently | [ ] no eye contact |

Figure 5.1

An excerpt from the "Patient Behavior" section of an SP script template. Excerpted from a training document from the University of Texas Medical Branch's resources for medical educators.

actors. For the medical educator who prepares it, the patient template takes the form of a lengthy document that can cover everything from demographic descriptors of the SP, like age, gender, race, sexuality, and physical description, to a detailed account of the SP's complaint, wardrobe, background, and affect.[18] For the patient, the details of the script can vary. The example shown in figure 5.1 is an excerpt from the "Patient Behavior" section of an SP script template. The section provides the range of dimensions for the patient's emotional state and how that state should present in body language and facial expression. The template is meant to program, as much as possible, the basic contours of the interaction, such that the responses of the physician-in-training can be judged both qualitatively and quantitatively against other trainees.

In turn, a template produces a script for SPs to follow. The "Back Pain Script" shown in figure 5.2 includes basic demographic details with added identifying information, including education, employment background,

Back Pain Script

(Nancy Owens, Age 44)

Chief Complaint: "I've had some back and leg pain and want help for it."

Identifying Data: College-educated accountant; workload stressful at times; married, one child, good home life.

Scenario: Your low back/left leg pain began about three months ago. You had a similar problem during the last few months of your pregnancy (your child was born seven years ago), but then none until three months ago. You consider yourself an athlete and can't run due to pain and intermittent numbness and tingling ("pins and needles" feeling) in the left leg.

Patient Profile: Concerned/anxious about this problem. You are in pain during the interview, but it is tolerable. Sitting is very uncomfortable, so shift around after several minutes. Bend forward slightly when sitting (put hands under knees—having knees higher than pelvis feels better). When walking, do so slowly with pelvis tilted forward. You have slow movements, with some stiffness in your back. Standing tolerance is 10–15 minutes. You bend over and rotate slowly. If asked to lie down, bring your knees up and flatten your back for comfort.

Figure 5.2
The beginning of an SP brief for a "back pain" patient, excerpted from a training document from the Baylor College of Medicine.

and possible stressors like feeling overworked. The "Scenario" and "Patient Profile" are written in the second person, "you." The Profile describes how the performance should unfold from the perspective of the performer. SPs are given a list of questions that they should expect to hear and the answers they should give to each. These range from questions about specific ailments and pains ("Q: Where is the pain? A: Lower back, especially left side and left leg") to questions about the patient's history, background, and typical behavior ("Q: Alcohol? A: Socially, one or two glasses of wine a week").[19] SPs are coached in how to embody these scripts and how to express their symptoms through their bodily comportment and reactions to physical examination. By memorizing, internalizing, embodying, and in turn exhibiting the details of their personae, SPs leverage their relationship to actual patients and create continuity between textbook knowledge, simulations, testing environments, and actual doctoring.

The scenario often begins with the trainee performing a standard history-taking. When they do so, the SP leverages an analogic relationship with actual illness through the performance of a life marked by experiences,

ITEM 17 – EMPATHY AND ACKNOWLEDGING PATIENT CUES

[5]	[4]	[3]	[2]	[1]
The interviewer uses supportive comments regarding the patient's emotions.	The interviewer uses NURS (name, understand, respect, support) or specific techniques for demonstrating empathy.	The interviewer is neutral, neither overly positive nor negative in demonstrating empathy.	No empathy is demonstrated.	The interviewer uses a negative emphasis or overly criticizes the patient.

Figure 5.3

Excerpt from the MIRS showing how to score a trainee on their demonstration of empathy. The scores run from 5 (highest) to 1 (lowest), and there are descriptions of how to score for each number on the scale.

routines, and behaviors that can be called up to corroborate their symptoms and to form a complete picture of a person in need of medical care. This performance of experience is meant to render their bodies as readable through a theater of transparency. In the words of Rachel Hall, the theater of transparency is a "mode in which the citizen's episodic affirmations of life and futurity are rehearsed, compelled, enacted, repeated, and confirmed."[20] The only difference here is that the citizens in question are stand-ins for the lives and futures of others.

Together, the actor, template, and script create a controllable, relatively repeatable scenario against which medical students can be evaluated. The same scenario can be repeated and used to test the variability in student knowledge, empathy, and skill. A final piece of paperwork is used to record judgments of the physicians-in-training: the Master Interviewing Rating Scale (MIRS) is one tool used to score up to twenty-seven features of the interaction, beginning with the trainee's introduction and progressing through such factors as how the trainee elicits information about the impact of the illness on the patient's family, the trainee's demonstrated empathy, and their acknowledgment of the patient's cues (figure 5.3).[21]

A standard uses the stuff of the local and the specific to code and compel behaviors in the future. The standardized patient program takes this logic to its extreme, using the bodies, voices, and physical responses of actors, the predictable and scriptable formulas of templates, and the ideology of psychometric testing to create an idealized version of how medical care ought to be delivered. The purpose of improving these interactions is not only to improve the bedside manner of would-be physicians, but also to improve the overall care of patients. Activist and patient movements to gain recognition for suffering, patient safety, and equity in care have all demonstrated that the emotional and affective tenor of medical care improves patient outcomes and a patient's willingness to access care in the future.

SIGNALING THE BODY IN PAIN

The standardization work involved in the SP program relies on a belief that we can communicate our pain or suffering through language and representation—that there are media for expressing and recognizing suffering. This is evident in the ways that the standardized patient program relies on semistructured encounters where patients share their symptoms and histories. But if the communication of pain were easy, then the standardized patient program wouldn't exist. In reality, the expression of one's pain and the recognition of that pain by others are both complex, interpersonal events that can be easily overwhelmed by social expectations, prejudices, and the noise of mediation.

In her groundbreaking work *The Body in Pain*, Elaine Scarry examines the difficulty of expressing pain and the obstacles that exist to finding genuine sympathy for another person's suffering. To be in pain, Scarry argues, is to have certainty, but to hear of another person's pain is to have doubt. This gap between certainty and doubt is further obstructed by the fact that pain often also inhibits expression, either because pain itself is debilitating or because it "resists verbal objectification."[22] Not only is another person's pain fundamentally inaccessible as a physical phenomenon, the ways that we normally bridge intersubjective gaps are not readily available in the case of pain. Language and other forms of symbolic representation do not convey either the intensity of pain or its felt immanence. Finally, Scarry argues that because pain is often ineffable and there are limited outlets for its representation, there are few ways of achieving political recognition for it.[23] The result is a circular reaffirmation of pain's loneliness—the person in pain is certain of their experience but lacks any recourse for gaining compassion and recognition.

For Scarry, the only path to validating another person's suffering is for that experience to be objectified (i.e., turned into a common object of analysis) and lifted into a world of shared symbols and representations in a manner that retains a definite reference to the human body. Pain needs media. Once materialized and tethered to a shared understanding of embodied experience—that is, mediated—pain can be acknowledged, validated, and attended to.[24] Any attempt to alleviate or remedy pain faces a series of gargantuan tasks: bridging the intersubjective gap between feeling pain and having that feeling understood, having pain validated through symbolic representation, and achieving political recognition that can provide the grounds for ongoing and future recognition and remedy. The capacity for someone to clear these hurdles is also encumbered by the uneven distribution of recognizability. Gendered, raced, and stigmatized forms of injury, disability, and pain face greater obstacles to recognition and acceptance and demonstrate that the gains in intersubjective understanding are always contingent.

> > >

Medical systems have many tools for trying to bridge the intersubjective gap between patient and physician. There are tools for quantifying pain, for narrating or describing it, and for gauging an appropriate clinical treatment for it. These tools vary by discipline, region, and practitioner, but each seeks to do what Scarry describes and to mediate between a person who is suffering and a person who may be able to alleviate their pain. Widely used instruments for describing and quantifying the subjective experience of pain ask patients to give a rating to their pain ("on a scale of 1 to 10") or to rely on a standardized vocabulary to characterize pain. The widely used McGill Pain Questionnaire dates from the 1970s, contemporaneous with the standardized patient program, and sorts descriptions of pain into three categories (sensory, affective, and evaluative) and rates each sensation on a scale of intensity. Ronald Melzack, who originally developed the questionnaire, provided it as a means for transforming the experience of pain into information that "can be treated statistically."[25] To put this in the terms that Scarry sets out, instruments like the McGill Pain Questionnaire can transmute the ambiguity of another's pain into a statistical regularity, placing the subjective experience of a stranger on a grid of patterned normativity.

Scarry's approach is a humanistic and philosophical one. But sharing one's pain and accepting another's pain are processes embedded in social and historical milieus. They are processes marked not only by an intersubjective gap, but by institutions built on norms and standards that accord different bodies more or less legitimacy in their expression of pain. Even with tools to represent and measure pain, there is inequity in the recognition of suffering.[26]

In 2016, researchers at the University of Virginia published a study in the *Proceedings of the National Academy of Sciences* entitled "Racial Bias in Pain Assessment and Treatment Recommendations, and False Beliefs about Biological Differences between Blacks and Whites." The authors show that among white-identified laypersons, medical students, and medical residents, there were widespread false beliefs about the experience of pain among racially coded Black patients and white ones. Among the laypersons, 73 percent ascribed to at least one false belief about the biological difference between Black and white people, including that "Blacks age more slowly than whites," "Black people's blood coagulates more quickly than whites'," and "Blacks' skin is thicker than whites'."[27] Among those with some medical training, 50 percent of their participants still ascribed to at least one false view. And among those who were medically trained, false beliefs about pain reception correlated with biases in their treatment recommendations.

In other words, among a statistically significant segment of the sample, medical students and residents (1) had false views about biological differences among Black and white people, (2) ascribed "fantastical" attributes to Black patients, and (3) were less likely to follow an accurate treatment recommendation for Black people suffering pain.[28] As the authors of the study note, their work takes place in an American medical system that has inherited reified racial differences that propagate both a false human racial taxonomy and a false view of Black bodies as being more resilient: "beliefs about biological differences between Blacks and whites—beliefs dating back to slavery—are associated with the perception that Black people feel less pain than white people, and also with inadequate treatment recommendations for Black patients' pain."[29]

Throughout the history of the United States, a belief in the racial peculiarities of white and Black people has been used as a justification for slavery, torture, and medical experimentation premised on a cultural view of the Black body as resilient and physically predisposed to withstand pain. Activist and social justice organizations like the Black Lives Matter movement have had to work to gain recognition for the particular suffering of Black bodies (hence the erasure of that specific pain expressed in the antislogan "All lives matter") within a system that forecloses the possibility of such recognition. Within white supremacist systems, the denial of pain suffered by people of color is a strategy to actively delegitimate the political demand to reckon with the history of colonial and settler exploitation and ongoing forms of acute structural violence, and a refusal that some kind of reparations ought to be paid.[30] As the University of Virginia study makes clear, this history has a subtler cost: the ongoing misrecognition of felt pain within medical establishments. A person seen as Black is less likely to receive necessary medical care or relief if their pain and suffering are not recognized.

One of the mechanisms of white supremacy is the way that it has paradoxically secured a position of dominance within institutions through the pretext of nonwhite resilience. The history of scientific and medical research in the Global North is marked by dehumanizing and eugenicist projects and devious experiments on the bodies of Indigenous people and people of color. From the moment of settler dispossession up through the twentieth century, as Harriet Washington documents in *Medical Apartheid*, the history of the United States is marked by medical experimentation on the bodies of minorities. Studies like the Tuskegee experiments—which allowed syphilis to develop untreated in the bodies of Black men, without their knowledge or consent, all in the name of research—are outrageous violations of personal sovereignty. But as Washington's history makes clear, the episodic nature of white outrage at such events masks the longstanding continuity that has entrenched medicine as a white-serving institution and bred a well-earned distrust of the profession among nonwhite patients.[31]

We must situate the widespread denial of Black pain among medical professionals as a continuation of this history, as a manifestation of its banality. We must also not ignore the fact that Black pain is not monolithic or itself a natural category of suffering. So-called Black pain is an intersectional and relational concept and a narrative of experience within institutions built on white supremacist precepts: women, mothers, the economically oppressed, Indigenous, queer, disabled, and gender nonconforming individuals are much more likely to be denied necessary medical care or to suffer nonconsensual medical interventions.[32] The denial of Black pain is a white supremacist excuse for the objectification of human bodies in medicine and science, and it erases human vulnerability to structural harm. Angelique Davis and Rose Ernst refer to this process as "racial gaslighting," by which they mean the "political, social, economic, and cultural process that perpetuates and normalizes a white supremacist reality through pathologizing those who resist."[33]

The denial of Black pain is part and parcel of a white supremacist society in which marked minorities are more likely to suffer harm, all while the harming entity simultaneously claims that the minority is less likely to feel pain. The most widely held false belief among the medical students studied by the researchers at the University of Virginia was the belief that Black bodies have thicker skin. It is a literal manifestation of the fantasy that Black bodies are naturally built to withstand structural and literal abuse. The fantastical capacity ascribed to Black bodies to withstand pain is dehumanizing twice over: they categorize a racialized body as superhuman (abnormal) while materially narrowing the options for remedying the actual harm that is suffered.

The standardized patient program was developed in the 1960s in a moment when physician attention to the affective dimensions of their work was an afterthought. It became an entrenched part of medical training as these dimensions, as well as their standardization, gained recognition as both socially and medically efficacious. But the program continues to operate within an institution that propagates widespread disparities in care and narratives about racialized bodies that deny their subjective experience of pain. As Scarry argues, any person seeking to communicate their pain faces a double bind—being certain of their own pain while responding to the doubt and skepticism of their audience. Overcoming this bind within the medical establishment is a process that relies on the standards of medical history-taking, symptom-reporting, and a duty of care. But the standard is not enforced equally. The credibility of another's pain is established through medical rituals formed within institutions built on discriminatory practices. The playacting of SPs needs to be appreciated within the context of racialized violence and the failure to accept some bodies in pain as legitimate.

NOT *NOT* SICK: THE SUSPENSION OF DISBELIEF IN STANDARDIZED PATIENT SCENARIOS

In Barrows and Abrahamson's first published description of the standardized patient program, they write:

> The concept of the "programmed patient" involves the simulation of disease by a *normal person* who is trained to assume and present, on examination, the history and neurological findings of an *actual patient* in the *manner of an actual patient*. This person is then used as the subject for clinical testing of student performance.[34]

What we see in this distillation of the standardized patient program is the way that normalcy is an imagined canvas for abnormality. Barrows and Abrahamson make a number of assertions about the feasibility of standardizing human performances and creating an emotionally charged exchange. The first is the notion that a person could be "programmed." The second is the suspension of disbelief that this work requires: suspension on the part of the actor *and* the physician-in-training. Barrows and Abrahamson address this suspension head on in their original description: "The student is informed that this is a simulation but that he is expected to treat the subject *as he* would a patient."[35] Encoded in the basic instructions is a productive fictionalism, in which all parties act *as if* it were the real thing in order to encode the promise of a better encounter in the future.

Finally, the simple declaration that the patient would be a normal person, trained to assume and present the characteristics of an actual patient, delineates a presupposition that there are "normal" and "diseased" people and that the former can learn to be like the latter with enough coaching. Erving Goffman describes "we and those who do not depart negatively from the particular expectations at issue" as "normals."[36] The standardized patient program appears to uphold such a distinction and to take Goffman's typology even one step further. Whereas "normals" are those people who can pass without stigma, the standardized patient program uses this apparent normalcy as the basis for creating a dependable and reproducible appearance of illness. Through the process of producing a new standard of care, the program reinscribes a binary division of human bodies as either normal (disease and disability free) or abnormal. In the attempt to codify an understanding of bodily difference, the program begins with a normate template of human well-being beyond which all maladies can be codified and pathologized. The normate, following Rosemary Garland Thomson, designates the "veiled subject position of cultural self, the figure outlined by the array of deviant others whose marked bodies shore up the normate's boundaries."[37]

Many standardized patient programs seek out actors whose bodies are not clearly marked by actual impairment or disability. In doing so, these putatively normate bodies become doubly significant: the healthy actor is always already capable of embodying the signifiers of illness through trained simulation (i.e., they are people), and they are free of the overdetermining noise of illness and disease (i.e., they are "normal"). Their power resides in a suspension of disbelief in a person's ability to shift to the simulation of impairment, illness, and disability without permanently embodying that social position—to live as the embodiment of the *as if.*

To put it in the terms of proxies and their maintenance, SPs can signify clean, hygienic versions of test data that can be used for the measurement of physician performance and that are preferred over the uncleansed and unpredictable embodiments of people living with actual illness, socially coded as disabled, or both. To understand these interactions and the use of actors in the simulation of illness and disability, we have to understand how medical institutions reinforce a system of compulsory and prototypical able-bodiedness and able-mindedness that relies on a medical and social understanding of disability.[38] To quote Lennard Davis, "the very concept of normalcy by which most people (by definition) shape their existence is in fact tied inexorably to the concept of disability, or rather, the concept of disability is a function of a concept of normalcy. Normalcy and disability are part of the same system."[39]

Disablement and impairment are relational experiences, contingent categories, and cultural artifacts that emerge from the interaction of bodies, identities, discourses, built environments, and the politics of medicine and state control.[40] Disability activists and scholars have often fought the medicalized descriptions of people's bodies that categorize, essentialize, and individualize them according to diagnostic criteria. The medical model of disability, it is argued, has historically been a tool of oppression and exclusion that is tied to forced institutionalization and a generalized view of "disability" as a medical problem and personal tragedy.[41] As a remedy to the medical model, many have advocated for a social model of disability proclaiming that disability ought not to be treated as an individualized ailment, but rather as a shared, social process of disabling. To put it simply, the medical model insists that disability resides in a person's body while the social model insists that the problem lies in society; the social model "is a deliberate attempt to shift attention away from the functional limitations of individuals with impairments onto the problems caused by disabling environments, barriers, and cultures."[42] The social model draws a distinction between "impairment" and "disability," wherein many people might have a range of impairments, but only some people will be categorized as "disabled." Hence, for most people in the Global North, moderate nearsightedness is not disabling because corrective lenses are widely available and there is little stigma about wearing glasses or contact lenses. On the other hand, many people with paraplegia will experience the world as disabling given the lack of accessible buildings and transit systems, widespread discrimination, and threadbare social welfare programs. The social model treats disablement as something that happens when society has failed to adapt to an impairment.

Although the social model has found purchase with many activists, academics, civil society, and some medical institutions, some argue that its distinction between impairment and disability is itself too rigid. By distinguishing between impairment as a description and disability as a social disadvantage, critics argue that the social model risks its own essentialization of impairment as an objective phenomenon that rises to the level of the social only when it becomes a disability.[43] Such a distinction risks arbitrary exclusions of people—for instance, those with chronic illness or chronic pain—for whom the label of "disability" might offer political potential and collective belonging.

As an alternative, Alison Kafer offers a political-relational model of disability as one that builds on social and minority understandings of disability. A political-relational model is attuned to the "failures and omissions of the built environment" that create more frictions for some embodied dispositions than for others.[44] But the political-relational model goes further—it's an attempt to pluralize the meanings and experiences of bodily instability, to unsettle the certainties of the medical and social models, and to refuse the depoliticization of disability. Kafer insists that disability must be understood as *relational*, meaning that a disability is experienced not in isolation, but rather through relationships: this includes the constitutive relationships that one has with the meaning of ability/disability, with family and friends, and with strangers.[45] Together, these relationships ripple out, and the social effects of stigma, discrimination, love, support, and compassion can fortify, twist, attenuate, and rend our shared networks of experience.

The standardized patient program is a technology built into the education and accreditation system of the medical profession and—more important—it is a crucial piece of the emotional infrastructure of medical work. It is also founded on the idea that illness, pain, and suffering enter the medical establishment through a relational encounter. The encounter is the meeting place, threshold, and exchange point between the world *out there,* as the messy world of personal, idiosyncratic experience, and the one *in here,* where knowledge is systematically organized. The SP encounter takes the form of an improvised though structured dialogue, through which one party comes to appreciate the nature of another's embodied experience and places it within a system of knowledge that recodes experience as diagnostic data. The entire encounter is held up by another form of relationality, the suspension of disbelief, which must be upheld by both parties, or else the whole interaction would capsize.

If another's pain is fundamentally unknowable then all empathy relies on some dimension of shared suspension: I trust you to relay your pain; and you trust me to believe you. The standardized patient program, then, is a form of relational training—a way of building up a chain of social relationships that might bring another person relief from suffering. It's a way to practice diagnosing by sifting through the information provided by the "patient," as well as a way to practice a willingness to believe, to trust another's experience as a valid report of their pain and suffering. While we might focus on the embodied experience of the actors in these scenarios, a political-relational approach to impairment and disability should equally attune us to the ways that the SP scenario is meant to entrain the disposition of a physician to the culturally unexpected experience of real compassion.

> > >

In *The Empathy Exams,* Leslie Jamison describes the kinds of maintenance and hygiene that she and her fellow SPs undertake between simulated examinations:

> Between encounters, we are given water, fruit, granola bars, and an endless supply of mints. We aren't supposed to exhaust the students with our bad breath and growling stomachs, the side effects of our actual bodies.[46]

SPs use these practices of upkeep to conceal and mask the parts of their bodies that would interfere with the encounter—body parts that would introduce noise into the media setting of the scene. This is a textbook way of controlling the variables in a testing scenario—of highlighting only the parts of the body and the emotional exchange that are supposed to be examined. It is also another example of the manual, embodied, and routine forms of data hygiene that workers in standardization processes use to maintain their test data. Instead of an ether-alcohol solution used to wash a kilogram or the folding/tearing of an illicit image examined in preceding chapters, the "data set" for an SP is an organic, living person. And while Jamison notes that she and her fellow participants are given water, food, and mints, the responsibility of upkeep falls to each actor to maintain their body as a canvas for testing the variability of physician performance.

Louise Aronson, a physician and educator, says that she has "often heard from students that the greatest challenge of standardized patient exercises was the suspension of disbelief. The most vociferous complained that circumstances so contrived couldn't possibly test them on actual practice skills."[47] As Aronson concludes, however, immersion in the scenario is possible for her—as an experienced physician—through the practiced execution of a kind of labor that she is accustomed to performing:

> As my own encounter began, I could see their point. Yet soon thereafter, with patient and student fully immersed in their roles, I couldn't help but behave—and respond—much as I would in so-called real life. There was artifice but also familiar work to be done.[48]

All attempts to build a standard around a proxy require a suspension of disbelief. Proxies are contrived—arbitrary, but precise—samples of the world that are provisionally treated as good enough stand-ins. They are makeshift, but they are necessary just the same. And so we suspend our incredulity to get our work done. With enough repetition, this suspension can take on the appearance of naturalness, or even objectivity. As Aronson says, the familiarity of a routine patient interview covers for the artifice of the proxy patients. The SP scenario needs to hold together just long enough for the necessary data to be gathered: the trainee determines an actor's history, and the medical educators evaluate the trainee's performance.

As consumers of media, we are accustomed to the generic conventions that establish suspensions of disbelief: the markers, frames, contexts, and affective investments that we bring to cultural texts to let ourselves be sutured into their artifice. In theatrical settings, the capacity to suspend disbelief is an ideal that must be upheld by both performers and audiences. In *An Actor Prepares*, Constantin Stanislavski quotes his director as describing "the test" of a successful performance as the "art of living a part."[49] This oft-repeated slogan of so-called method acting refers to the ways that an actor embodies an experience as theatrical reality. Within a scene, the emotional experience is real even if the scenario itself is contrived. It is possible, however, to see this technique of living a part, of suspending disbelief, as a theatrical trick to achieving and expressing verisimilitude, and to view it as a larger cultural technique for producing and verifying knowledge. It is also possible to see it as constitutive of the very practice of being a doctor.

In a popular manual for physicians, titled *Proper Doctoring*, published in 1984, near the end of his career, the British cardiologist David Mendel writes, "Although a sound knowledge of the facts is essential, a good doctor differs from a bad doctor more by his attitude and craftsmanship than by his knowledge."[50] Mendel's tract is composed of a series of doctoring precepts that he compiled throughout his career. The focus of his advocacy, however, is the "much-maligned bedside manner."[51] For Mendel, a properly honed bedside manner, and by extension a properly honed doctor, are a result of training and practice in the art of *playing* a doctor. Under the section "The Need for Role-Playing in Medicine," he writes:

> Doctors vary in the extent to which they act. Some ham it up. Others are to the manner born, and practice their art without artifice. Some confine their role-playing to their work; others never stop acting. Doctors range from the saintly and perceptive who do not need to play-act, through to the horrid and insensitive, who may neither realize, nor care, how beneficial a role could be.[52]

Mendel lays out a method, echoing Stanislavski's, for routinely inhabiting the role of "doctor" until the role feels natural. In this, his analogies are mainly theatrical. For instance, he writes in the section on "Rehearsing your role":

> The role of doctor, like the role of Hamlet, is not one which you can leap onto the stage and perform. In order to play Hamlet, as distinct from watching him, which anyone can do, you have to examine each word, each phrase, and take it all in the context of the whole play. . . . So you go on stage, knowing your role, knowing what you want to achieve—the perfectly treated patient— and you have to use a lot of learned techniques to achieve that end.[53]

What Mendel sees in the practice of proper doctoring is not a distinction between performance and objectivity, but the weaving of the two: the necessary acceptance that one's diagnostic is mediated through the relational encounter with a patient. The interface of the exam is always theatrical, and for both patient and doctor, its repertoire is learned through the coding of practiced gestures and speech.

>>>

In reenactment, there is often a fear that the theatrical rendition of an event will be either too real or not real enough. In the case of the former, there is the danger that reenactors will forget that they are involved in a practice of make-believe; in the latter, there is the danger that they won't give themselves over to the performance, that it will seem phony. Rebecca Schneider describes the friction of this relationship as a "queasy portal," through which performers can momentarily forget their embodied time and place. In her interviews with historical reenactors, Schneider says, "The differences or the *lack* of differences between faux and real might not necessarily be failures or threats."[54] There is no guaranteed meaning tied to the failure to delineate reality from reenactment; there is no strict hierarchy between "real" and "pretend" to which performers aspire. Instead, Schneider says, there are moments "when some things, like reading, *or even modes of critical thinking or patterns of analysis*, become habit-memory, they are skills, fully learned, available to call up as research tools or artistic craft."[55] For reenactors of the US Civil War, this form of habit-memory can be a way of encoding the traumatic memory of the past through performance. These are ways of animating collective, cultural forms of remembrance that can be recalled through the scripting of gesture.

Reenactment can also be a way of warding off an unwanted reality. Schneider's reenactors perform historical battles to reanimate a remembered moment in a controllable way. For many of her informants, performance is not (or not only) a proxy for a desired past, but a proxy for an unwanted future: by performing *now*, in this way, we will know how it felt *then*, and we will be able to prevent its recurrence. Performance as proxy, then, can act as a historical prophylactic, averting historical recurrence through an affective awareness written in gesture, costume, and real-time recreation.[56] But, it is, of course, both: Civil War reenactors might claim that their practice helps to prevent the recurrence of another bloody war, while in reality, they spark nostalgia for an antebellum, slave-owning South—calling forward the very resentments and violence that fueled the war in the first place.

SPs are not *reenactors* in the same sense as those who portray Civil War battles. There is no famous diagnosis of gangrene or Tourette syndrome to which actor and physician aspire. Instead, they play at a more general *enactment,* through a pursuit of a likeness to a generic or typical medical interaction. To do this, however, they need to walk the same line as historical reenactors. The roles that medical students and actors play in SP encounters operate in an agreed-upon realm of make-believe in which they are both "not actors" and "not *not* actors," what Richard Schechner describes as the "liminal realm of double negativity that precisely locates the process of theatrical characterization."[57] The unpredictability of human interaction is both the reason for the use of SPs and the reason that their status as fixed points is inherently unstable.

Critics of SPs claim that too much emphasis is put on simulation in medical education at the cost of experience with "actual patients," which threatens to make "simulation doctors" instead of actual ones;[58] and advocates of SPs claim that actual patients are too idiosyncratic and vulnerable, while computer simulations are too inert.[59] The implication is that SPs exist in a hierarchy of experiential value, somewhere below actual patients and above computer simulations. For the training and evaluation of physicians, however, the hierarchy is reversed: computer simulations would give the greatest amount of reproducibility across student encounters, and real patients the least. SPs, therefore, are a compromise between the rival commitments to verisimilitude and reproducibility that all proxies must negotiate.

Like some Civil War reenactors, however, simulated patients have a prophylactic function. Their use in the training of physicians is meant to achieve a standard of care that is both medically efficacious and compassionate. Just as some of Schneider's performers cite the embodied experience of war reenactment as a kind of memory-making in defense of repetition, SPs create a buffer between trainee physicians and the more vulnerable bodies of actual patients. An awkward touch, an inappropriate question, or an incredulous tone of voice—SPs must intercept these faults and blunders of incipient doctors, including all the gestures and attitudes of miscaring that can produce discomfort or manifest as a disbelief in another's pain and suffering.

>>>

Suspension of disbelief would not be necessary, however, if students were interacting with people who were actually experiencing the symptoms they described. Yet while patient interaction is a crucial part of a physician's training, it is not a viable means of *comparing* students. In SP scenarios, the embodiment of a simulated disease is a trained technique as well as a method for creating standardized tests. Simulation is a means of modeling behavior that always serves two purposes: to train individuals and groups in embodied habit-memories, while still providing officials with information and observations from a controlled test. Standardization works by scaling up from a controlled test to an actual implementation.

According to Barrows's original explanation of his program, from 1968, SPs offered many advantages over actual sick people and cadavers, the only two alternatives that were suggested at that time (the text that follows is paraphrased except where quoted). SPs are preferable because:

- They save embarrassment.
- The patient does not fatigue.
- "All necessary aspects of disease complications and prognosis can be freely discussed in front of the simulated patients without concern for their reaction to such information."
- The diagnosis is invented by the examiners before the examination can begin, so a controlled evaluation of the student is possible.
- The same clinical problem can be repeated to many examinees.
- The trained SP can report "objectively on the student physician's skills."[60]

By 1993, at a point when the standardized patient program was growing in popularity and Barrows was less defensive about its virtues, he gave these same reasons: SPs were ready and available whenever an instructor needed a patient, they could be used repeatedly, and they save the patient from mistreatment (but he doesn't specify what costs there might be to the mistreatment of SPs). But in 1993, Barrows added an interesting argument in favor of SPs: "The standardized patient is prepared for students to perform inadequately and is prepared to be used as a teaching and assessment tool. You have no concern about the student's making inappropriate remarks in the teaching situation or using poor examination techniques."[61]

Barrows was addressing fellow physicians, so the shift to the second person is telling, as it comes at the exact moment when he is highlighting the instrumentalization of patient bodies in terms that reflect their status as living people with sensitive bodies. "They" (in contrast to "you") are prepared to assume the status of a tool. This instrumentalization, this lack of concern about inappropriate remarks or poor examination techniques, is entirely a consequence of transforming living people into workable proxies. The process relies on the consistent and reliable reuse of SPs, but achieved without the erasure of what makes each person's body an idiosyncratic entanglement of experiences, memories, and feelings. It cannot work without suspending disbelief at the irreducible difference between one body and another.

In her ethnography of SPs, Taylor describes the authority of "presence" that the SP achieves by being both "real" and a "simulation":

> Simulation generally is distinguished by the premise that suffering is not present—it may be *there*, or *then*, but it is *not here-and-now*, and that is what makes the SP performance a "safe" learning environment for the student. At the same time, however, everyone I have spoken with agrees that the presence of the SP as an *actual person* is crucial to these performances, and is what distinguishes them from other forms of "mere" simulation.[62]

These are the paradoxes of proxies that are at once meant to be stand-ins for real-world situations and yet still be usable. Actual embarrassment or actual pain would make the simulation untestable or unreproducible, but what is being tested, in some ways, is the ability of a physician-in-training to manage the patient's emotional response to a diagnosis. SPs must be trained to react in ways that would imply the possibility of feelings like embarrassment, shock, or shame, while guaranteeing that those feelings will only be performances.

As Tobin Siebers discusses, the performance of disability by nondisabled actors is a way of sanitizing lived experience for a potentially uncomfortable audience by reassuring them that the person is really all right. It also rerenders the disability as a performance, turning a person's lived experience into signifiers of virtuosic impersonation: "the audience perceives the disabled body as a sign of the acting abilities of the performer—the more disabled the character, the greater the ability of the actor."[63] By disarticulating bodiness (being human) from bodily particularity, practitioners claim that SPs save both patient and student the embarrassment of vulnerability—the vulnerability of sharing one's pain and the vulnerability of being a novice professional. But they also treat disability as superficial—something to be worn—and an experience that can be made transparently accessible.

The standardized patient program, using the logic of testability, operates with a logic of prototypical able-bodiedness. Like the prototypical whiteness of test images, which treat white skin as the default and nonwhite skin as a special case, the program begins with the "able" and "healthy" body as the unmarked default condition, upon which illness and disability can be layered. For medical educators, the widespread use of able-bodied actors to perform illness and disability works to foreclose vulnerability and to make the entire scenario a low-stakes affair. It also mirrors a widely discredited technique for building sympathy for disability by having able-bodied people try to navigate the world blindfolded, in a wheelchair, or otherwise impaired.[64] For SP educators, the low stakes of disability masquerade and simulation open up possibilities for using SPs to combat other problems. This is because, as Barrows long argued, SPs are transitional aids; bridging the world of simulation with the world of lived reality. The nature of this transition can be varied. It can mean the honing of skills and knowledge, but it can also mean the recalibration of empathy and prejudice.

In one study in Germany, involving thirteen SPs and 200 medical students, researchers had the SPs wear "obesity simulation suits."[65] The aims of the study were twofold: to elicit from the students accurate body mass estimates; and to measure the prevalence of antifat stigma. By the authors' own metrics, the study was a success. Not only did the students accurately predict the range of body mass index scores, the encounter demonstrated that there were widespread antifat attitudes. The masquerade became a teachable moment, free of real vulnerability. The obesity suit, overencoding the body of an SP, transformed a real medical stigma into sanitized data. The authors describe the strengths and limitations of the study as follows: "The study was conducted only in a simulated environment and not with real patients with obesity."[66] The argument implicit here is that no vulnerable people were harmed in the production of the data, though vulnerability and harm were modeled and measured. The standardized patient program became a kind of machine, able to model and simulate prejudice, and while doing so create usable data. But to do so, embodiment must be turned into spectacle.

A number of standardized patient programs have begun to make an active effort to incorporate people living with disabilities into their corps of SP actors. At the Tufts University School of Medicine, for example, medical educators employ patients with disabilities as SPs and equip them with, for instance, a complaint about shoulder pain. Shoulder pain is specifically chosen for its banality: "It is not only a common problem for all adults, but it has particular implications for patients with physical disabilities, as many are critically dependent upon shoulder function."[67] In designing the scenario, the educators chose not to inform the students that their SPs would have a disability. In their reasoning, "the interaction would be more realistic, and they (the students) would have an opportunity to grapple with their own reactions, including their biases, in a setting that closely resembles actual practice."[68] What is remarkable about the Tufts case is not only the use of SPs with disabilities, but that the "reveal" in the scenario plays on the expectation that typical SP encounters will not use actors living with a disability. Like the obesity study in Germany, it uses the encounter as a provocation, to elicit stigma and reflection. Unlike the obesity study, it relies on both the vulnerability of people living with disabilities and the potential for embarrassment on the part of students.

There is a risk in dismissing the theatrical elements of knowledge production and the ways that knowledge is secured through ritual as somehow giving lie to the process. That is to say that the purpose of unearthing proxies and their rituals is not to highlight moments of fantasy and make-believe as inherently deficient; the fact that proxies are in widespread use does not mean that we can unlock the contrivance of scientific and technological meaning-making by simply showing proxies to be artificial. It is true that these systems rely on contrivances, but that is not what makes them susceptible to critique. There is no "mere" theater. Instead, there is only a deeply interwoven set of techniques of embodiment and technologies of memory that tie together action, word, setting, and practice to form a context in which things can be made discernable, measurable, and judged.

MODELS, PATIENTS, AND MODEL PATIENTS

Barrows's first SP was a twenty-two-year-old woman named Rose McWilliams, whom he met through USC's Art Department, where she worked as an artist's model. According to Barrows, McWilliams was coached to "have a paraplegia, bilateral Babinskis, dissociated sensory loss, and a blind eye."[69] She was soon joined by a second SP, Lynn Taylor, who was another model. Initial responses to the standardized patient program were harsh; few physicians outside of USC thought that the project was viable. The program was maligned for its proximity to Hollywood, its reliance on a vapid form of performance, and the artifice of simulating disease.[70] These criticisms were often attached, in particular, to the women working as SPs, whose well-honed work as professional performers brought them into the medical establishment as programmable patients.

In 1965, in the early stages of the program, the Associated Press learned of Barrows and Abrahamson's work and syndicated a story in several papers. In the *San Francisco Chronicle,* the headline read "Models Who Imitate Patients: Paradise for Medical Students" (figure 5.4); in the *Los Angeles Herald Examiner,* "Hollywood Invades USC Medical School"; and in *Newsday,* simply "Model Patients" (figure 5.5 shows a drawing and caption from the

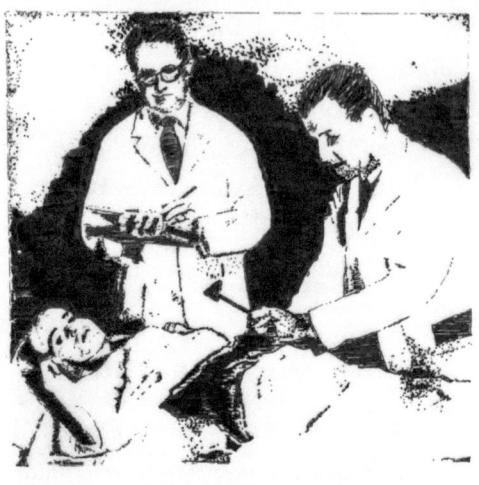

Models Who Imitate Patients

Paradise for Medical Students

THE EASY WAY TO LEARN MEDICINE
Rose McWilliams, Professor Howard Barrows and student John Goodman

Figure 5.4

An artist's interpretation of a headline, image, and caption from a *San Francisco Chronicle* article (September 28, 1965). The caption reads, "The Easy Way to Learn Medicine." The image portrays one of the original standardized patients, Rose McWilliams, with Howard Barrows and one of his medical students. Image: R. R. Mulvin.

story). The story, as it was rewritten from a wire report, always began the same way: "Scantily clad models are making life a little more interesting for University of Southern California medical students," and then went on to describe Taylor as "a shapely brunette" and McWilliams as "the blonde."[71] SPs echo, therefore, the pattern that the history of test images set forth: women's sexualized bodies are folded into a technical bureaucracy and a technique of knowledge formation, which both renders them as objects of analysis and erases their labor as stand-ins.

METHOD MEDICINE. Rose McWilliams, actress and model, performs as a patient with neurological troubles during a checkup by medical student John Goodman, left. Dr. Howard S. Barrows, associate professor of neurology in the University of California School of Medicine, is observer.

Figure 5.5

An artist's interpretation of an image and caption from a *Newsday* article (September 28, 1965) about SPs, also showing McWilliams, Barrows, and Goodman. Image: R. R. Mulvin.

Like the *Playboy* that was supposedly lying around a digital image processing lab (discussed in chapter 3), ingenuity is understood through the instrumentalization of a female model's body. The origins of the Lena image are recounted as the confluence of boredom, innovation, and happenstance. This is repeated in the account of SPs and the way that having models ready-at-hand at USC (on a campus famous for its ties to the entertainment industry) was narrated as a fortunate coincidence. In neither case do the makers of these standards take credit for seeking out women's bodies as templates. Instead, the disavowal is in part what allows this process to reoccur throughout history and across disciplines. If accepting responsibility for using women as test instruments threatens in some way the technical process that those bodies enabled and provokes a confrontation with the forces of desire, sexualization, or even an unjust power differential, then the originators need some way to shore up the technical aspects of their choice as a means of distancing themselves from the cultural implications.

By all evidence, the Associated Press account was not the story that Barrows sought to tell about his new program. He saw the use of the art department as a canny strategy to harness existing university resources. In a handbook that he produced for interested educators in 1971, Barrows reaffirmed his commitment to this method: "If I were to start a program elsewhere, my first move would be to inquire at the local drama department or a local amateur or professional acting society as to whether or not there would be potential interest [in being an SP]."[72] He is quick to deny the claim that his program is unduly indebted to its proximity to Hollywood. He says that during a sabbatical year at McMaster University, in Hamilton, the "Steel Mill Capital of Canada," he relied on "a housewife" who "became one of the most effective and versatile simulators with whom I have worked."[73]

Barrows's advice to others hoping to establish their own standardized patient programs is to seek out a facility for SP work wherever it could be located: "Do not ever ignore interested technicians, secretaries, housewives, and so on. Motivation is the real key."[74] These suggestions expose an implicit gendering of SP labor. The instructions, with a certain level of frankness, are addressed to medical professionals whose expertise attunes them to "housewives" and "secretaries" who might have interest in SP work. Just as, in the history of the Lena image, a *Playboy* centerfold was ready-at-hand to become a test image, the women in Barrows's world are a ready set of model patients, already stand-ins. Meanwhile, all the images of USC medical students in the early days of the program depict men, and the early models were all identified as women. Although contemporary standardized patient programs seek out a more diverse cohort of actors, the origins of the program are thrown into sharp relief: women possess an immanent instrumentality that can be used by physicians (men) and educators (also men) who should be ready to take advantage of it.

As the standardized patient program grew, the range of people who would become actors also expanded. SP work now lends itself to a wide array of potential actors, as physicians need training across a broad range of bodies, demographic attributes, and personality types. But the labor of being an SP is also often most amenable to a worker with already precarious or intermittent work. Who can be a proxy is partly determined by who has the time and needs the money to learn the techniques of embodied illness and disease. Theater departments and local theater companies are a ready resource for acting ability, but as Barrows's list of other possible actors— "technicians, scientists, housewives"—indicates, financial need is also a good motivation, and flexible working hours are an added bonus.

What Barrows foresaw was that SP jobs could be part of a person's pool of possible "gig work" assignments. In his 1993 summary of the benefits of the standardized patient program, he wrote that SPs were preferable to real patients because they were always available ("Unlike real patients, the standardized patient can be available at any time and available in any setting") and their service is transactional ("The standardized patient is paid to be examined *over and over again* by numerous students").[75] The so-called gig economy is often ascribed to the twenty-first century (in the Global North) and the deindustrialization of labor forces, in which an increasing number of people piece together a living through many part-time, nonpermanent positions and full-time employment becomes more rare (along with benefits like a pension, disability and parental leave, and sick time). SP work in the 1960s and 1970s was imagined as part of a person's piece-work assemblage of jobs, and SPs are drawn from pools of people just making do. For professional or amateur actors, SP work is a supplement to their intermittent work and often described as a way of honing and sharpening their performance skills.[76] For everyone else, SP work is a way of instrumentalizing something that everyone has: a body that is always ready to be medicalized.

The work of SPs is vital to how physicians are trained, in a standardized way, and licensed as professionals. Their work is infrastructural and enacted through their bodies and performances, employing their status as living people, able to stand in for others. To understand what it means to per-form infrastructural labor as an SP is to understand the ways that the work is rewarded, is regarded as a form of employment, and (possibly) reflects a relationship to notions of professionalization and expertise. Rose McWilliams and Lynn Taylor, the first SPs, were paid 45 dollars an hour in 1965 (approxi-mately 370 dollars an hour in 2020). According to the reporting at the time, this was the fee that they normally charged for modeling work.[77] This pay treated their professional skills as models as transferrable to their work as model patients and as equally worthy of remuneration. Today, SPs make an average of 15 dollars an hour, though they can earn as little as 7 dollars and as much as 25 dollars.[78] In general, more experience does not lead to more pay. However, invasive exams do earn more—on average 40 dollars an hour—though one report notes that working as an SP for breast and vaginal exams can earn as much as 55 dollars an hour because these patients "are also instructing students how to perform them."[79]

Figure 5.6 shows a fee schedule from the University of South Florida, readily obtained from the Health Sciences Center's website.[80] By looking at the entire one-page document, we can see an encapsulation of the stan-dardized patient program, how the program is presented to prospective SPs, and the ways that it presents a hierarchy of expertise, which follows an axis of bodily invasiveness. The form begins with a title that is imme-diately asterisked: in reading the heading, potential SPs should also note that all their encounters may be observed and/or audio- or video-recorded.

STANDARDIZED PATIENT ACTIVITIES AND PAY SCHEDULE+,**
UNIVERSITY OF SOUTH FLORIDA HEALTH SCIENCES CENTER

Interviewing only
Students will learn or be evaluated on communication skills, interpersonal skills, and history taking. SPs will either "play" themselves (no training) or take on a role (requires training). No physical examination will be involved. Activities may be one-on-one with the student or in a small group setting.
- Interviewing only $7.00/hr

Target students: Year 1 and 2—medical *(Physical Diagnosis-PD)*

Physical Examination only (no breast, pelvic, or rectal exams)
Students will learn or be evaluated on physical examination skills. SPs will be in patient gowns and students will perform the physical exam by touching the skin. Training is required when there are particular elements of the physical that a SP must record and/or when there are symptoms that must be simulated.
- Physical Examination only $10.00/hr
- With breast exam $5.00/hr (additional)

Target students: Year 1 and 2—medical *(Profession of Medicine, and PD)*

Focused History and Physical (no pelvic or rectal exams)
Students will learn or be evaluated on their interpersonal skills and their ability to obtain a history and perform a physical examination that revolves around a specific medical complaint (cough, fatigue, etc.). The students generally spend less than 15 minutes at a time with the SP. SPs are almost always trained to a specific medical script and physical presentation.
- Requires training and student evaluation $15.00/hr

Target students: Year 3—medical *(Clinical Performance Examination-CPX)*

Comprehensive History and Physical (no pelvic or rectal exams)
Students will be evaluated on their ability to obtain a comprehensive history and physical examination. Generally, SPs will also evaluate the students' interpersonal and communication skills, as well as examination technique and sequencing. The students generally spend at least 30 minutes, and up to 1 hour, with the SP.
- Requires training and student evaluation $20.00/hr
- Does not require training or student evaluation $15.00/hr
- With breast exam $5.00/hr (additional)

Target students: Year 2 and 4—medical *(PD and Videotaped History and Physical Program)*

Special Physical Examination Consultants
Students will learn examination techniques from faculty clinicians and then perform the exam themselves. On occasion, the SP will function as a model for evaluation of student technique by a chaperone.
- Female breast examination (no SP teaching or student evaluation) $15.00/hr
- Pelvic examination $35.00/hr
- Male rectal and prostate examination; including penile and testicular examination $35.00/hr

Target students: Year 2 and 3—medical *(PD and Clinical Clerkships)*

Training
- Training (payment for training will be discussed prior to beginning the programs and is subject to the discretion of the Program Coordinator) $10.00/hr

+All encounters with students may be observed directly or video/audiotaped—a consent form is required.
**All pay rates are subject to change at the discretion of the Office of Curriculum and Medical Education.

Figure 5.6

Fee schedule for SP work from the University of South Florida Health Sciences Center.

Observing, and potentially recording, SP encounters are key elements of making testable interactions.

The remainder of the form proceeds through an increasingly intimate menu of possible interactions. It begins with "Interviewing only," in which SPs may be able to "play" (put in scare quotes in the original) themselves or a character, and then proceeds to "Physical Examination Only" and "Focused History and Physical," which combines the first two stages while homing in on a particular ailment, and then continues to "Comprehensive History and Physical," which lasts up to an hour. The list ends with "Special Physical Examination Consultants," which is an opportunity for SPs willing to train in receiving breast, pelvic, and prostate examination (including penile and testicular examination).

The ladder of possible encounters includes a great deal of nested information. At each rung, the document notes whether the SP will require special training or preparation to participate. For instance, at the "Focused History and Physical" level, SPs may be trained to act out a specific ailment, and at the next stage, SPs can be asked to evaluate students in their "interpersonal and communication" skills. Each level also increases in hourly pay as SPs are asked to do more, be more trained, or permit greater amounts of physical touch and bodily investigation. SPs at the entry level are paid 7 dollars an hour, but SPs at the comprehensive level can be paid 20 dollars an hour if they are prepared to evaluate the students, and an additional 5 dollars an hour if they will permit a breast exam.

This document demonstrates how the line between "make-believe" and the actual lived boundary of a person's body must be inscribed in the management of proxies. It shows how an SP is both not a patient and not *not* a patient. The more invasive examinations are accorded more pay not only because of their potential effects on the actor's body, but also because of the expertise that this invasiveness elicits—an expertise tied to living and sensing in that actual body. Regular SPs are coached not to give physicians-in-training any advice during the examination, since it's beyond the scope of their assignment and any advice they gave would taint the exam results. But SPs who opt for invasive examinations are an exception; they are often expected to coach students in practice scenarios because they are the only ones who can be sure if invasive procedures "feel right." Expertise and bodily sovereignty are aligned in this case, which only emphasizes how much they can be severed in the case of typical SP scenarios.

In the process of legitimating standardized patient programs, the work of simulating disease and disability has been transformed from a professional skill (that of modeling) to a low-wage, part-time job, with little to no prospects for a raise or advancement, and one in which a person's value as a resource is tied to their bodily affordances and trainability as a performer. Temple University's guide for aspiring SPs notes that most of their actors earn less than 2000 dollars per year from their work.[81] And Temple's program is the rare case where guidance is provided.

As SP education expanded and gained legitimacy, it needed to do so through the logic of standardization. The standardized patient program needed to be reliable, verifiable, and reproducible, which meant that SPs could be only one node in the evaluation system. By treating SPs as amateurs, the process also minimizes their role in evaluating physicians. When McWilliams and Taylor first began as SPs, Barrows gave them the responsibility of recording what transpired with the student. While SPs still complete questionnaires after their interactions to record their experiences, the weight of the evaluation falls on educators, through one of a few standardized evaluation templates.[82] Some standardized patients have also demanded more input in evaluating student physicians on the grounds that their subjective responses to trainee physicians, as well as their implied resemblance to *any* potential patient, give them special authority to speak for patients more generally.[83] Meanwhile, standardized patient program coordinators must intervene when SPs appear to overstep their roles, like when an actor provides "feedback that seems to tread into the domain of medical expertise."[84] SPs who demand more say in the evaluation of students are negotiating for greater power over their own representation in a standardization process.

When bodies are treated as proxies of physical and affective measurement, standardization will necessarily require political negotiation over who has the power to access and describe those measurements. SP demands for more input in the evaluation of trainees must be folded back into the knowledge infrastructures of the medical system, in which the cooperation of these actors is necessary to maintain a coherent pedagogical apparatus. As Greg Downey writes, "workers formally or informally engaged in complex information labors . . . are continually re-embedded in each new round of the knowledge infrastructure."[85] These particular workers hold tremendous power because of their capacity to leverage their real bodies to become proxies for anyone's potential maladies.

IMAGINING NORMAL

In a training video produced by Barrows in 1988, he demonstrates the steps involved in developing an SP encounter. The video, entitled *Acute Paralysis of Both Legs in a Young Woman*, shows how a physician should take a patient's history and conduct a physical exam.[86] For the performer, Barrows describes the development of the case, the dress rehearsal, and the final encounter. He instructs the young woman about how to simulate paralysis. When he pokes her leg with a sharp pin and asks her, "Do you feel that?" she is expected to say "Feel what?" instead of "No." The "No" would betray the artifice of the scenario—an indication that she is suppressing her actual sensation for the purpose of simulation. Instead, verisimilitude is achieved through a total denial of sensation: "Feel what?"

It is telling that Barrows would use paralysis (and specifically numbness) as a template for training SPs, as it clearly demonstrates the bifurcated identities that SPs must occupy when their bodies are both scripted and lived in: the SP feels the pinprick but must comport herself as if she does not. This is the embodiment of *as if*, acting as a present surrogate for an otherwise absent Other. At the end of his interaction with the actor, Barrows simply asks, "Can you conceive of all of this happening to you?"[87] Barrows's question is a final attempt to make sure that the act of imagination that the SP participates in will accord with the experiential reality of the encounter.

By breaking the world into manageable chunks, standard makers must make certain decisions about how best to represent the world. These choices about the best way to represent the world for the given task take the form of proxies, which are used to build and maintain standards and have real effects on their representational and distributive capacities. SPs have the potential to irritate the smooth production of a standard of care on the very terms of this proxy logic. A person cannot be a simple, manageable chunk of the world, and SP encounters are highly scripted because their manageability is threatened by the possibilities of improvisation and the idiosyncrasies of a person's body and temperament. Moreover, SPs betray the always-performative separation of the world *out there* and the world *in here*. While they are employed as supposedly normal people play-acting as ill, they also have their own, unique, leaky bodies that breach the membrane between the testing scenario and the outside world. In the history of proxies, the viability of a standard is endangered when its representational capacities are called into question or when the functional separation of inside and outside breaks down. SPs are special because they work only through their capacity to bridge both inside *and* outside.

The standardized patient program is partly a technique for helping train doctors to see patients as more than a history of symptoms, and instead to engage them as agents of their own care. Standards operate through normative conceptions of how the world ought to operate. The hope for more humane medical care or of more empathetic medical professionals is an optimistic fantasy that needs to be encoded in new standards and norms. As these norms, standards, and procedures play out, they will do so, partially, through the bodies of SPs who are trained and practiced in the art and technique of remembering a history that isn't theirs, possessing symptoms that aren't theirs, and responding to the probing questions and hands of doctors-to-be in ways that are sympathetic to the many and varied patients *out there*. SPs have an impossible but necessary task to stand in as proxies for all of us. Although the standardized patient program relies on a suspension of disbelief, it is based on a certain fact: everyone will need care at some point in their lives, and everyone deserves empathy and recognition of their pain.

6 CANNED CHANCE: METHODS FOR FOLLOWING INFRASTRUCTURE

Stories about proxies are full of acts of concealment, erasure, and disavowal. The stories in this book are suffused with attempts to mask the arbitrary decisions and protocols that make standardization possible and make infrastructures persist. They are also full of stories of the spectacularity of proxies: people observed the bombing of Yodaville from grandstands and highly aestheticized images of these trainings circulated online; the International Prototype Kilogram (IPK) was consecrated in a ritualistic burial; the Lena image became an icon, its own proper piece of auratic media, far beyond its original purpose; and standardized patients have been used frequently in film, television, and literature to mock the peculiarity of medical education. But the spectacularity of proxies contributes to their concealment by affirming that these things, these places, and these people are special—they are vested with the power to stand in for the world. They are our chosen delegates.

To varying degrees, the concealment of standards is what makes them ordinary, what helps a standard integrate with infrastructure, and what allows infrastructure to operate unperturbed. The study of proxies, then, is the study of the basic conditions of infrastructure, the labor that it takes to make and maintain infrastructural conditions, and the process of repairing and recuperating those conditions when they inevitably break down.[1] This approach to proxies, standards, and infrastructures provokes a necessary question: what tools or techniques are available to uncover this process? Or, to put it bluntly: how do we make infrastructures into objects of analysis, and what is at stake if we do?

In this final chapter, I want to sketch out some of the ways people have turned infrastructures into objects of analysis and suggest how a theory of proxies, as well as a process for tracing proxies, can supplement these existing strategies. As a coda to the preceding chapters, I want to offer that researching and writing the biographies of proxies can constitute a technique for getting at the lifeworlds and poetics of infrastructural thinking. In approaching infrastructure in this way, I want to propose three complementary ideas: first, that the concealment of infrastructure (including standards) is an ongoing and always partial process; second, that transparency is a political claim, not an ontological status of infrastructure; and, third, that based on the context, various tools and techniques can be used to map the relational and political dynamics of infrastructure. Following from these ideas, I explore two prominent and conspicuous techniques for mapping infrastructures and standards—the ironic repurposing of standardization in artistic practice and the revelatory promises and techniques of critical infrastructure studies—both of which rely on a promise that they can expose the taken-for-granted.

It is "a joke about the Meter," responded Marcel Duchamp when asked to comment on his artwork *3 stoppages étalon* (3 standard stoppages), shown in figure 6.1.[2] In this work, Duchamp measured three meter-length pieces of string, dropped them onto a canvas from 1 meter above, and traced the twisted and sloping lines they created. These outlines reconfigured a measurement unit, the meter, as three slouchy lines, and Duchamp used these shapes as templates for three drafting straightedges, the objects displayed in *3 stoppages étalon*. In his *Box of 1914,* Duchamp included fragmentary comments about the artwork, describing it as a distortion of standardized measurements:

The Idea of the Fabrication

–If a straight horizontal thread one meter long falls from a height of one
     meter onto a horizontal plane distorting itself *as it pleases* and creates a
     new shape of the measure of length.

> −3 patterns obtained in more or less similar conditions: *considered in their relation to one another* they are an *approximate reconstitution* of the measure of length.
>
> −The 3 *standard stoppages* are the meter diminished.[3]

Duchamp later named this process "canned chance."[4] In fact, *3 stoppages étalon* is representative of a familiar trope from the interwoven histories of standardization, measurement, and artistic experimentation. If Duchamp's *3 stoppages étalon* is a joke about the meter, what's the joke about? First, the International Prototype Meter was the basis of all length measurements in the metric system when Duchamp created his artwork. It was, by all appearances, the opposite of a piece of string dropped "as it pleases." Instead, the Meter was a piece of platinum-iridium forged to be durable and stable—to mutate as little as it pleased. All proxies live in things—metal, paper, pixels, or flesh—and the exchange of a metal alloy for household string is a joke about the durability of proxies, a joke about the theater of objectivity, and

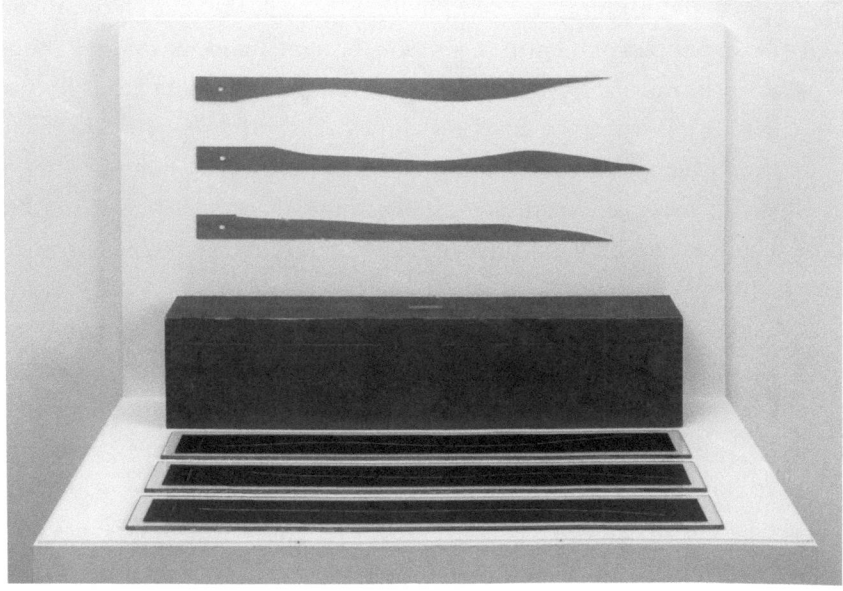

Figure 6.1

Marcel Duchamp, *3 stoppages étalon* (3 standard stoppages) 1913–1914, replica from 1964. ©Association Marcel Duchamp/ADAGP, Paris and DACS, London 2020. Photo: ©Tate.

a joke about the fantasy of authority attached to sturdy stuff. Making the Meter out of platinum-iridium invokes a fantasy of timelessness and the notion that the materiality of a proxy will live up to its (supposedly) invariant reference points.

Standards are, to play off of Duchamp's terminology, canned decisions about how to represent the world in a limited fashion; standards put a lid on a lengthy and necessarily cultural practice of choosing workable segments of the world. The official Meter, the subject of Duchamp's joke, was based on a fraction of the Earth's meridian, compiled during seven years of painstaking measurement of the French landscape during the peak of post-Revolutionary turmoil.[5] The International Prototype Meter, like its sibling the IPK, was a materialization of the highest aspirations of Enlightenment ideals. Accordingly, Duchamp's second joke is at the expense of French Enlightenment idealists, who could just as easily have saved the seven years of trouble by dropping a piece of string on a canvas. Standards, as this book describes, are enacted through strict protocols and rituals for creating systems of communication and coordination through the sharing of proxies. Proxies, as samples of the world *out there*, become stand-ins for how things might work. But *3 stoppages étalon* turns the process on its head. Instead of choosing a proxy as a fixed point and establishing a repertoire to maintain it, Duchamp created a small-scale ritual to maintain a random set of points. The joke is on the theater of objectivity: there would be no need to shore up the legitimacy of a proxy if it were never intended to be legitimate in the first place.

Duchamp's last joke concerns the fact that standards use traceable references to maintain order and create usable systems. The metric system works with very few errors because its references are accessible, reproducible, and guaranteed through a network of linkages that can be traced to the system's bureaucratic home in France. The nested and internetworked dependability of standards is what makes them infrastructural.[6] That is to say that standards work as the conditions of operability for other systems (e.g., the standards that make up networked computing allow applications to work across devices; the standards that make up the highway system allow any street-legal car to use it). Standards use fixed points so that each subsequent operation built on that standard doesn't need to reperform the labor of fixing points. If every time someone wanted to build a house, they had to get the architects, engineers, contractors, and inspectors to agree on length standards, building would take a lot longer. In fact, this was the case at one time. Prior to standardized units of length, traveling masons involved in cathedral construction in the twelfth century used adjustable templates to format their tools to each town's local measurements using, among other things, pieces of string reminiscent of Duchamp's stoppages.[7]

In *3 stoppages étalon*, Duchamp's final product is a series of three straightedges that likewise are templates: representational technologies that materialize a process in a physical artifact. Like the standards that the artwork is intended to lampoon, *3 stoppages étalon* uses a ritual to create its proxies. Although each Meter-proxy captures a different moment in the fall of the string onto the canvas, and their differences to one another demonstrate the ridiculousness of the exercise, they nonetheless evoke the ways that standards operate as "recipes for reality":[8] begin with a method for fixing points; use those points to create proxies; and use those proxies to create new objects, systems, and expressions. Thus, *3 stoppages étalon* is a useful way to expose the logic and idealism of standardization. Among its many jokes, the artwork prods the performative and ritualistic labor that goes into transforming the arbitrariness of standardization into the appearance of objectivity.

Duchamp was not the first (and far from the last) artist to use the tools of standardization and infrastructuralization in his work. Such tools make attractive subject matter for artists because they embody a simplified formalism in which representation is reduced to elementary lines, shapes, and colors. The inclusion of common measurement tools in creative works also can be a way of signaling to a spectator a common way of seeing things. In Renaissance painting, artists were trained to use a familiar repertoire of stock objects that would allow viewers to gauge how accurately the artist had reproduced a particular perspective: "A painter who left traces of such analysis in his painting was leaving cues his public was well equipped to pick up."[9] These stock objects varied by locality but often relied on the knowledge of common containers: barrels, sacks, and bales. In this way, they foreshadowed contemporary proxies like the Lena image, which can be used to benchmark individual talent and the efficiency of an algorithmic process, or other, standard three-dimensional (3D) digital objects like the Stanford bunny and the Utah Teapot.[10] The Utah Teapot, also known as the "Newell Teapot," is a 3D representation of a Melitta teapot—a sample of domestic hardware meant as a stock object for testing a program or a user's abilities, and easily recognized by other users. It was one of the first of such objects, and it frequently appears in computer-animated media as a winking nod to knowing insiders.[11] While Duchamp could turn to the metric system and tweak its arbitrariness with aleatory play, and Renaissance painters could invoke standard objects to cue perspectival awareness, contemporary art is similarly festooned with standard measurement tools. Contemporary artists frequently demonstrate their familiarity with more recent common items used in the testing and calibration of codecs, file formats, and standards.

In her video installation *How Not to Be Seen: A Fucking Didactic Educational .MOV File,* the artist Hito Steyerl begins with a handheld resolution target, a test image, and poses holding it in a fashion reminiscent of the Shirley cards discussed in chapter 3.[12] The video proceeds through other resolution targets used by the US government and military in the calibration of spy satellites since the 1950s. The image in figure 6.2 depicts one of these resolution targets, a tri-bar test image in the California desert. The targets have fallen into disrepair, with grass and trees sprouting through the concrete—a visual reminder of the porousness of proxies. Steyerl's work ties together several of the threads discussed throughout this book. It brings us full circle from the visual and experiential stand-in of Yodaville, presenting yet another materialization of the military imaginary and its apparatuses for controlling space through surveillance. The fact that the targets are themselves objects of fascination for contemporary artists also foregrounds the theatrical and mediatic potential of proxies, which always exist as unsettled instruments—at once operating as objects of measurement and props in a theater of objectivity.

The April 2006 cover of *ArtForum* (figure 6.3) features a photograph by the artist Christopher Williams—a reproduction of a Kodak Shirley card.

Figure 6.2
A tri-bar resolution target from the 1950s, used to calibrate satellite imaging technology. The target is located near Cuddeback Lake, in the Mojave Desert, California. Photo: ©CLUI. Courtesy of the Center for Land Use Interpretation.

The artwork, with the unwieldy title *Kodak Three Point Reflection Guide © 1968 Eastman Kodak Company, 1968 (Meiko laughing), Vancouver, B.C., April 6, 2005,* is a staged reenactment of the now-familiar genre of the test image. The Museum of Modern Art describes the work as "[sending] up the aesthetic conventions of photographic representation." Like Duchamp's *3 stoppages étalon,* this work is meant to transform the "invisible" standard of representation into an aestheticized object and turn the format of photography into an object of critical appraisal.[13]

In 2005, Julie Buck and Karin Segal mounted an exhibit at Harvard's Sert Gallery titled "Girls on Film," which displayed film control strips, including many so-called China Girl images (the show was written up in the *Harvard Gazette* under the headline "A Bevy of Unknown Beauties").[14] For the 2012 Whitney Biennale, the artist Lucy Raven contributed a much-heralded series of works titled "Standard Evaluation Materials," which included a set of film projector calibration images and speaker test tones, accompanied by a public lecture about contemporary test films used in digital projector calibration.[15] The infamous phrase discussed in chapter 3 and used by the Kodak executive to describe new film stock that could

Figure 6.3
An artist's interpretation of the April 2006 cover of *ArtForum* magazine, featuring a representation of a Kodak test image, or Shirley card. Image: R. R. Mulvin.

"photograph the details of a dark horse in low light" (coded language, as Lorna Roth argues, for the film stock's capacity for representing the skin of people of color) became the title of a photography exhibition by the artists Adam Broomberg and Oliver Chanarin. In staging this exhibition in Toronto, they mounted Shirley cards on billboards throughout the city.[16] As Robin Lynch argues, the interplay of technological appropriations of art and artistic appropriations of technical materials can be read in some ways as an implicit attempt to bridge the two cultures that supposedly divide the human and the scientific. However, any such attempt is never guaranteed; it is subject to the power dynamics of representation, the gendering of technical disciplines, the instrumentalization of women's bodies, and the use of racialized prototypes.[17]

These are just the beginning: proxies, standardized instruments, and the application of tacit knowledge of commonplace objects are recurrent themes in the history of art and attest to the inseparability of all three.[18] Like Duchamp tweaking the metric system's performative objectivity, the use of a proxy as the basis of artistic expression is nothing short of a rite of passage for these materials. The Lena test image has not only appeared in HBO's *Silicon Valley,* as discussed in chapter 4, but has also been the subject of poems, reenactments, and video art. Artists like Trevor Paglen and Luke DuBois have used different versions of the image and the centerfold in recent years, with DuBois even receiving a cease-and-desist letter from *Playboy* for his artwork.

If there is an overemphasis on visual proxies in this book, that shouldn't be taken as an indication of their exclusive role in artistic appropriation. Test sounds also are featured frequently in artwork; for instance, the Suzanne Vega song "Tom's Diner," which was used in many tests of what would become the MP3, became material for artwork by Ryan Maguire.[19] Standardized patients are also a popular topic in television, film, and journalism, appearing as comic relief in episodes of both *Seinfeld* and *ER*. Moreover, standard-makers are often eager to reappropriate their own proxies. The first standardized patient, Lynn Taylor, was painted by Phyllis Barrows, the wife of Howard Barrows (she was apparently in a class where Taylor modeled). Later, in 1997, Howard Barrows used the painting in a special issue of the medical journal *Caduceus,* on a topic called "Simulation in Medical Education."[20] The examples here demonstrate the fluid movement of proxies between different domains of cultural expression.[21]

Proxies are not simple instruments used in the standardization process: rather, they are cultural artifacts used by people steeped in culture, and through reuse, proxies become new materials for creative expression. The dynamic interaction of creative and technical expression is never unidirectional, and this demonstrates the impossible task of fully separating the technical life of proxies from their affective, creative, and embodied dimensions. As the Lena image indicates, and examples like "Tom's Diner" corroborate, standard-makers are not isolated from the culture at large. They are workers, and consumers, embedded in their own cultural milieus and their proxies are suffused with the cultural work of analogy and adhesion: they allow us to think through things to make new connections.

One of the goals of this book is to show how the nodes in the circuit of culture (identity, production, representation, regulation, consumption) could be expanded to include standardization, which is interlinked with every one of these other nodes. [22] The identities, consumption practices, and practices of representation of the people who work in the standardization process bear directly on the standards they create. Examples of this include the suturing of soft-core porn into the basic visual architecture of the internet and the scripting of embodied performance of disease and disability into the standardized patient program. This is not a novelty of the twentieth century. Standards have always soaked up the world around them. They tell stories about how to represent the world and in turn crystallize an abstracted version of the world that they set out to represent. Throughout this book, I have relied on an artist to represent certain artifacts of the world surrounding proxies: diagrams, images, video stills, and journal and magazine covers have each been reproduced and reinterpreted by hand. These works are meant to underline the contingent connection of proxies to their claims of verisimilitude—or the *realishness* of proxies. In addition, the illustrations are meant to foreground the manual and embodied labor of making and maintaining proxies that lies at the heart of the histories in this book.

A turn toward materiality within media and infrastructure studies has often implied a turn away from the cultural and representational—as well as the assertion or implication that we can study one without the other.[23] As the histories of proxies demonstrate, however, even the most basic standards and infrastructures, including the measurement standards of the metric system, are always built through cultural, representational, and embodied practices. They are thoroughly material and technological, but these practices never escape the human work of bringing the world into representational comparison. Artistic practices can provide an avenue for exposing the logics and inner workings of standards and infrastructures; but they also highlight the flows of texts, images, and cultural detritus that circulate more widely, and they expose any strict division between the cultural, the technological, and the scientific as always fictional.

THE TRANSPARENCY HYPOTHESIS

Where artists use ironic recontextualization to tweak the deadpan serious-ness of standards and infrastructures, scholars of infrastructure have often used the inevitable breakdown of all technologies to expose how such sys-tems work. The argument goes as follows: whereas infrastructure is typically a ground on which other things figure, a breakdown turns that ground into a figure for analysis. Within a growing body of work at the intersections of standards, infrastructure, media, and technology, scholars use this widely accepted premise to expose the givenness of standards, infrastructures, norms, platforms, and other related phenomena. This work depends on a transparency hypothesis to argue that these objects of study deserve atten-tion in part because they are taken for granted.

The most complete articulations of the transparency hypothesis are put forward by Geoffrey Bowker and Susan Leigh Star and their collabora-tors. Bowker's 1994 introduction of the very popular and useful method of "infrastructural inversion,"[24] which tasked researchers with looking to the embedded infrastructural preconditions that make science and technology possible, was followed in 1996 by Star and Karen Ruhleder's now-canonical definition of infrastructure.[25] Their definition is used by many scholars focused on the problematics of infrastructure, and it is often invoked as the basis for a new kind of "infrastructure studies."[26] Among the most cited and repeated features of infrastructure that Star and Ruhleder defined is the manner in which infrastructure is embedded, invisible, and revealed only when it breaks down. They write:

- *Embeddedness.* Infrastructure is "sunk" into, inside of, other structures, social arrangements and technologies;
- *Transparency.* Infrastructure is transparent to use, in the sense that it does not have to be reinvented each time or assembled for each task, but invis-ibly supports those tasks; [. . .]
- *Learned as part of membership.* The taken-for-grantedness of artifacts and organizational arrangements is a sine qua non of membership in a com-munity of practice. [. . .]
- *Becomes visible upon breakdown.* The normally invisible quality of work-ing infrastructure becomes visible when it breaks; the server is down, the

bridge washes out, there is a power blackout. Even when there are back-up mechanisms or procedures, their existence further highlights the now-visible infrastructure.[27]

These features of infrastructure are recapitulated regularly wherever infrastructure is said to "exist in the background,"[28] where "science and media become transparent when scientists and society at large forget many of the norms and standards they are heeding,"[29] where platforms are described as "whatever the programmer takes for granted when developing,"[30] and where comparisons are made between the relative transparency of infrastructure to other conditions of social life: "software code . . . is much less visible than law."[31] There are many more examples that repeat this familiar, powerful refrain: like breathing warm air onto a cold window, we might convey some shape of ignored, overlooked, and unseen objects. This is a powerful gesture, and one that accorded some measure of legitimacy to a nascent, interdisciplinary field.

The transparency hypothesis works in concert with a turn toward materiality and an analytic rematerialization of information technologies that were mistakenly or misleadingly described as immaterial for a generation or more. In massive, networked, digital communication systems, where control and resistance are exerted at the level of infrastructure, a realignment of investigatory interests is especially necessary. A new alignment should not, as the argument goes, focus exclusively on the surface content of communication technologies and media, but rather on the conditions that make communication technologies and media possible. Such work is conscientious in its unearthing and exposing of standards, infrastructures, and technologies; many of these objects *were* ignored and *were*, practically speaking, invisible to the kinds of investigations that were normally regarded as legitimate.

Another complication comes from the contradictory meaning of "transparent" in different contexts. In infrastructure studies, it has come to mean that something escapes attention, is ready-at-hand, or can be taken for granted. In institutional discourses (like those of the state, the corporation, or the university), "transparency" has the opposite meaning: it is a name for the exposure of processes, data, and practices that are otherwise considered to be hidden; or it can be the claim that particular processes, like those surrounding Big Data, artificial intelligence (AI), or behavioral tracking, will expose otherwise-hidden laws of social life. Calls for greater transparency are often calls for exposition, for making the secret public. But we all know from experience in these institutions that claims of offering greater transparency from political, corporate, or university actors are often attempts to control and manage oversight of their operations. Limited transparency, as much as total concealment, can be a tool of obfuscation. Hence, promises from tech companies that they will make their decision-making processes more transparent are an attempt to immunize themselves against other claims made against their business and labor practices.[32]

Likewise, in recent years, researchers have used "fairness, accountability, and transparency" within AI and machine learning as benchmarks of algorithmic in/justice—though these terms too have been coopted by the very institutions they were meant to question. As Mikkel Flyverbom argues, these negotiations over transparency and concealment are not zero-sum struggles, but rather a part of a larger social practice of "managing visibilities." In managing visibility, organizations wield transparency as a resource to be employed in negotiations over power and control, and in the formation of organizational identities.[33] What this leads us to conclude is that transparency is a political claim, a discursive resource, and a relational tool that gets instrumentalized in the struggle over boundaries and claims for justice and accountability.

Instead of seeking to expose the transparent, this book approaches standards and infrastructures as technologies that undergo constant, yet always partial, concealment. Concealment is treated here as a repetitive and reparative process. This is most obvious in the ways that the IPK underwent cleaning, and more subtle in the ways that image engineers responded to criticism of the Lena image from fellow students and coworkers. If, by definition, infrastructure is invisible—with *infra* referring to its occupation of a level below the readily perceptible—there is perhaps no good to be gained from upending a solid etymology. Yet, as Lampland and Star stipulate, it is "*good* infrastructure, by definition [that is] is invisible."[34] And here we should pause in order not to let the word "good" slip by because a distinction between what is seen as good infrastructure and what is seen as bad infrastructure ought to pique our interest. The experience and appreciation of infrastructure will vary by any number of sociodemographic conditions, including location, age, ability, and access to alternative systems. For thirty years, there has been an "infrastructure crisis" in the Global North.[35] If we are in a constant state of crisis about our basic infrastructure, then is any of it good, and by extension, invisible?

The term "infrastructure" as a category of critical analysis and as a category of social experience was always meant to be relational.[36] Critically gazing at the flows and barriers of infrastructure ought to prioritize the perspectival and social aspects of what makes a way of doing things seem transparent for some people at some times, and impossible for other people at other times.[37] As Kregg Hetherington puts it, the transparency of infrastructure is dependent on a "geography of uneven development."[38] This is clear for anyone who can't take infrastructure for granted, for whom breakdown is a fact of life, or anyone who lives in a place that is singled out as a target of neocolonial development.[39] As Gabrielle Hecht shows us, infrastructures have uneven "valuation/waste dialectics" that permit the making of modernity's scales.[40] We must understand the ideology of infrastructure-as-taken-for-granted as the dream of frictionlessness, the dream of handiness, and the dream of development.[41] This means that we have to account for the ways that infrastructure's promise is an always unfulfilled assurance of modernity, that things can always get better.

Transparency has politics. Sometimes things are hidden from view because they are repugnant or unjust.[42] Sometimes they are hidden because revealing them would expose an underlying exploitation. Drawing together the transparency of the clear glass skyscrapers of Toronto's modernist architecture with Canadian government claims to truth and openness, Jas Rault argues that settler colonialism operates through colonial "tricks of transparency," and an "architecture that means to feel like a natural, inevitable and inescapable environment."[43] Armond Towns argues that the manner in which infrastructures are discussed as taken for granted throws into relief how Black bodies are treated as always ready-at-hand resources and, in being treated as such, rendered invisible.[44] Rachel Hall teaches us how to view airport security through a theater of transparency, showing that it is unevenly enforced on racialized bodies.[45] For some, transparency is a threat; for others, it is a promise that one might be able to take the conditions of their lived environment for granted. Infrastructure is never only a system of substrates; it is always a means of distributing and redistributing security, access, recognition, and the conditions of making do.

SEEKING TRACES

As this book has described, things, places, images, and people undergo many forms of maintenance, repair, and transformation to allow them to continue functioning in their capacity as the material and representational underpinnings of infrastructure. All these acts of erasure, concealment, and suspended disbelief are relational, provisional, and contingent, and they are meant to manage the visibility of knowledge-making practices.[46] Think of the keepers of the IPK washing away the cylinder's contamination and the way that they consecrated the IPK by burying it with paperwork; think of the ripping or tearing of the Lena image and the stories that those engineers told for how and why they ripped or tore or folded the image *just so*; think of the standardized patients suppressing their actual bodies (via breath mints, granola bars, and other means) to highlight their simulated diseases and disabilities; think of the doctors playacting until it seems all too real because their careers depend on it. Proxies are crucial to the making of standards and infrastructures, and they too operate only within organizational practices of managed visibility.

So, how should we study things that are meant to be invisible if we cannot or will not wait for their breakdown-induced inversion? The artist, scholar, and writer Ingrid Burrington provides a handbook for seeing and understanding the internet infrastructure in New York City, which she describes as "hiding in plain sight."[47] In *Networks of New York*, she has created a field guide for deciphering the insider codes of urban architecture, including the cable markings, manhole covers, surveillance cameras, and antennae that saturate a city landscape but mostly go without notice. For Burrington, uncovering the black box of technology does not require special access, but rather a way of seeing and a particular positionality:

> The trick of how to see the internet isn't tech know-how or gaining access to secret rooms. It's learning what to look for and how to look for it. Learning how to see and pay attention to the fragmented indicators and nodes of networks on any city street is also a process of learning how to see and live within a world full of large, complicated systems.[48]

Using hand-drawn sketches, Burrington's brilliant field guide turns the opaque internetworks of concealed infrastructure into type-images like those of a bird-watching or tree-spotting guide. It converts the impenetrability of infrastructural semiosis into a pocket-size glossary of infrastructure's plumage, all without the need for breakdown. Her approach bridges the artistic representation of standards and infrastructure and the excavatory impulse of infrastructure studies.

If artistic appropriation and infrastructural inversion are techniques for defamiliarizing technologies that are taken for granted, then this book builds on these techniques. Proxies and their protocols (including data hygienics, instructions for performance, declarations about their accordance with the natural world, and so on) need constant attention and that attention leaves a trace. Through the process of writing a history of a proxy it is possible to follow how things, places, and people are turned into representations of the world and function as the necessary fictions of knowledge systems. To study infrastructure through its material assemblages, protocols, performances, documents, and embodiments is to understand how infrastructure works and is made to drift from view. But this book has aimed for something else, too: to account for what is excised, what is made a remainder of the process of creating a proxy. It has done so as a matter of justice and equity and as a matter of exposing the representational and world-making limitations of infrastructures.

As standard-makers expanded the ambit of phenomena that they might standardize, they were both driven and accompanied by a widening scope of things that seemed measurable. Thus, this book traces a filament of standardization throughout the long twentieth century, from basic measurement units to the standardization of digital images to, finally, the attempt to standardize medical care through the use of human actors employed as test patients. These histories both add to an understanding of how standards and knowledge infrastructures developed and give us an understanding of who works to make standards possible. These cases were also chosen because they left behind a body of evidence that tells us something about how proxies are chosen and maintained through the routines, rituals, and embodied performances of scientific and technical labor.

My research took me to the sites of official standardization and its histories: the Library of Congress, National Archives, National Library of Medicine, and National Institute of Standards and Technology; but it also led me to the less official places where the traces of standardization reside: countless hours spent reading digitized reports about image processing, following the message board discussions of standardized patients, navigating to photography sites where people exchange test and calibration images, and fruitlessly trying to get permission to see the IPK. It involved a dozen different borrowers' cards, a documentary regime that could attest to my identity,[49] and at each, a different set of embodied techniques for creating, requesting, or making scans and photocopies to produce my own set of traces and linkages between objects, people, and places. These are part of the embodied experiences and routines of archival work.

Where Carolyn Steedman describes the "dust of an archive" and the specific form of possession that haunts the historian, we might add to that dust the sad sandwich or premade burrito that we lug through humid Washington on our way to College Park, or the eerie loneliness of the picnic bench at the National Library of Medicine.[50] The burrito of the archive is the topic for another day, but these practices extend the larger world of proxies and their traces to mundane routines and protocols, the forms of access we can or cannot acquire, and the makeshift infrastructural labor of research.

>>>

Figure 6.4
The author in the process of creating a standard stoppage. String and pen on cardboard.
Photo: Lewis Bush.

Proxies are clarifying: their creation, maintenance, and use illuminate the connective tissue between the people, objects, and protocols that make infrastructure vibrant. The process of following proxies centers on the cultures that surround technical and scientific knowledge production. It trains a gaze, or a perspective, on the articulation work that everyone must perform.

In the final days of this project, I found myself re-creating *3 stoppages étalon*. I stood with Lewis Bush in his backyard and tried to drop some string onto pieces of cardboard that Lewis had lying around his studio (see figures 6.4 and 6.5). The string was twisted from being stored in a ball, and the wind carried the string away from the cardboard. It took many attempts. I have to confess that I had to move much closer than a 1-meter height to get a usable shape. It was a reminder of the contours of proxies: the embodied labor, the communities of practice, the makeshift materiality, the haphazard interventions of place and circumstance.

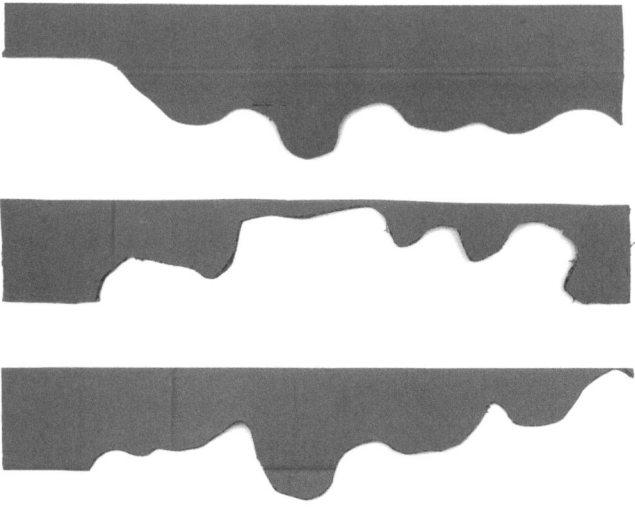

Figure 6.5

"Three More Standard Stoppages." Cardboard. Dylan Mulvin and Lewis Bush (2020). Image: Lewis Bush.

Proxies are real things, real places, real people; but they are also always memories of a world gone by and forecasts of a world to come. While they appear durable, they are porous, flexible, and, finally breakable. The proxies in this book are not unique, but they are stickier than some, having picked up the traces of the places where they've traveled.[51] The cases given here map a way of studying stand-ins as the pragmatic and practical artifacts for representing the world and the assumptions that are made of that world.

This book began with a simple question: to whom or to what do we delegate the power to stand in for the world? Some of the answers to this question have included cities in the American Midwest, shipping containers in Arizona, pieces of metal in France, centerfolds in Los Angeles, and medical actors anywhere. But these were only starting points. A concern for the cultural labor of standing in also demands that we ask: To whom do we delegate the power to *choose* stand-ins? And, finally, *for whom* do these delegates stand in? These final questions, which underlie the histories throughout this book, attune the analysis to the conditions in which proxy work happens, the injustices of representation and performance, and the labor politics of standing in. Our references shape who we are, how we think, how we communicate, and how we build shared worlds. It is no surprise that when these references take the form of recognizable objects, places, or people, they are often enchanting. When viewed from afar, or askew, the particular enchantment of, say, a piece of platinum iridium can appear ridiculous. Or, seen from another standpoint, the power to choose a centerfold to act as a stand-in for the world of images loses its aura. Instead, it appears as the manifestation of the constitutive injustices of visual culture.

Proxies are in operation in all places where communities of practice share representations of the world to form a common understanding. So let this be an invitation to you to take stock of the common reference points of your own knowledge infrastructures: the examples, prototypes, and stand-ins of a world *out there*. Our shared references may be the material through which we think and make new connections, but they are never permanent. The cultural work of standing in takes form in the "ritualized repetition of norms" that produce and stabilize difference, and encode it in the materiality of culture.[52] Our identities, our bodies, our worlds are formed through the ways that norms persist, and the labor of creating and maintaining technology is one way that norms are encoded in the infrastructure of everyday life. Our proxies are cultural, and at the same time as they shape the infrastructures through which culture circulates, they will be reshaped by the ebbs and flows of power and the demands for more just representation. The purpose of cataloging these proxies is to show how practices of standing in, as well as the work of embodiment, representation, and memory, impinge directly on the capacity for people to navigate institutions and achieve recognition within systems not of their own making.

Notes

CHAPTER 1

1. Hans Vaihinger, *The Philosophy of "As If,"* trans. C. K. Ogden (New York: Harcourt Brace, 1924), 93.

2. Russell W. Glenn, Jody Jacobs, Brian Nichiporuk, Christopher Paul, Barbara Raymond, Randall Steeb, and Harry J. Thei, *Preparing for the Proven Inevitable: An Urban Operations Training Strategy for America's Joint Force* (Santa Monica, CA: RAND Corporation, 2006); Geoff Manaugh, "Yodaville," *BLDGBLOG*, December 6, 2015, http://www.bldgblog.com /2015/12/yodaville/.

3. Mark Bowden, *Black Hawk Down: A Story of Modern War* (New York: Signet, 1999); *Black Hawk Down*, dir. Ridley Scott (Los Angeles: Columbia Pictures, 2001).

4. The report was commissioned by the Office of the Secretary of Defense and US Joint Forces Command.

5. Glenn et al., *Preparing*, xv (emphasis added).

6. If we think of Yodaville as an instrument in the exercise of empire, it is redoubled when we acknowledge that the Yuma Proving Ground's proximity to the US-Mexico border is a bulwark of the US occupation of that highly militarized boundary. As Audra Simpson and Lisa Ford have both argued, the layering of rationality onto a practice of make-believe is central to the logic of settler colonialism. By treating the nation and its borders as a rational and legal entity, colonial powers retrospectively justify the dispossession of territory. Audra Simpson, "The Ruse of Consent and the Anatomy of 'Refusal': Cases from Indigenous North America and Australia," *Postcolonial Studies* 20, no.1 (2017): 18–33; Lisa Ford, *Settler Sovereignty: Jurisdiction and Indigenous People in America and Australia 1786–1836* (Cambridge, MA: Harvard University Press, 2010).

7. Geoffrey C. Bowker, *Memory Practices in the Sciences* (Cambridge, MA: MIT Press, 2005).

8. The territory designations are taken from the map available at www.native-land.ca. As the site states, these designations are not meant as official/legal boundaries. Indeed, as mapmaking was crucial to European colonization and occupation, the history of boundary-drawing

is an extension of that colonial project. Here, the named territories and the reservation are offered as context for the specific emplacement of Yodaville within the history of US imperial occupation.

9. Sara Ahmed, *Strange Encounters: Embodied Others in Post-Coloniality* (London: Routledge, 2000), 132.

10. Madeline Akrich. "The De-Scription of Technical Objects," in *Shaping Technology/Building Society: Studies in Sociotechnical Change*, eds. Wiebe E. Bijker and John Law (Cambridge, MA: MIT Press, 1992), 205–224; Michel Callon, "Society in the Making: The Study of Technology as a Tool for Sociological Analysis," in *The Social Construction of Technological Systems: New Directions in the Sociology and History of Technology*, eds. Wiebe E. Bijker, Thomas P. Hughes, and Trevor Pinch (Cambridge, MA: MIT Press, 1987), 83–103.

11. Greg Downey, "Virtual Webs, Physical Technologies, and Hidden Workers: The Spaces of Labor in Information Internetworks," *Technology and Culture* 42, no. 2 (2001): 209–235; Paul Edwards, *A Vast Machine: Computer Models, Climate Data, and the Politics of Global Warming* (Cambridge, MA: MIT Press, 2010).

12. Marilyn Strathern, *Reproducing the Future: Anthropology, Kinship, and the New Reproductive Technologies* (London: Routledge, 1992), 33.

13. Andrew L. Russell, *Open Standards and the Digital Age: History, Ideology, and Networks* (Cambridge: Cambridge University Press, 2014), 16.

14. Paul Du Gay, Stuart Hall, Linda Janes, Anders Koed Madsen, Hugh Mackay, and Keith Nagus, *Doing Cultural Studies: The Story of the Sony Walkman* (London: SAGE, 2013).

15. Mary Douglas, *How Institutions Think* (Syracuse, NY: Syracuse University Press, 1986).

16. I have simplified the causality here. Humans never act, for Latour, out of a simple desire to use technology to achieve a single, unalloyed goal, but rather in a nonreducible network of relationships with other humans and nonhumans. See Bruno Latour, "Mixing Humans and Nonhumans Together: The Sociology of a Door-Closer," *Social Problems* 35, no. 3 (1988): 298–310; Bruno Latour, "On Technical Mediation," *Common Knowledge* 3, no. 2 (1994): 29–64.

17. Latour, again: "The speedbump is not made of matter, ultimately; it is full of engineers and chancellors and lawmakers, commingling their wills and their story lines with those of gravel, concrete, paint, and standard calculations." Latour, "On Technical Mediation," 41.

18. Langdon Winner, "Do Artifacts Have Politics?" in *The Whale and the Reactor* (Chicago: University of Chicago Press, 1986), 19–39; Trevor J. Pinch and Wiebe E. Bijker, "The Social Construction of Facts and Artefacts: Or How the Sociology of Science and the Sociology of Technology Might Benefit Each Other," *Social Studies of Science* 14, no. 3 (1984): 399–441.

19. Virginia Eubanks, *Automating Inequality* (New York: St. Martin's Press, 2018); Sasha Costanza-Chock, *Design Justice: Community-Led Practices to Build the Worlds We Need* (Cambridge, MA: MIT Press, 2020); Ifeoma Ajunwa, Kate Crawford, and Jason Schultz,

"Limitless Worker Surveillance," *California Law Review* 105, no. 3 (2017): 735–776; Frank Pasquale, *The Black Box Society: The Secret Algorithms That Control Money and Information* (Cambridge, MA: Harvard University Press, 2015); Stop LAPD Spying Coalition, "Location-based Policing: New LAPD Technologies, Same Racisms," September 5, 2019, https://www.citywatchla.com/index.php/2016-01-01-13-17-00/los-angeles/18377-location-based-policing-new-lapd-technologies-same-racisms.

20. Lorraine Daston and Peter Galison, *Objectivity* (Brooklyn: Zone Books, 2007), 21–22.

21. Robert Kohler, *Lords of the Fly: Drosophila Genetics and the Experimental Life* (Chicago: University of Chicago Press, 1994); Karen Ann Rader, *Making Mice: Standardizing Animals for American Biomedical Research, 1900–1955* (Princeton, NJ: Princeton University Press, 2004); Nathan Ensmenger, "Is Chess the Drosophila of Artificial Intelligence? A Social History of an Algorithm," *Social Studies of Science* 42, no. 1 (2012): 5–30; Greg Siegel *Forensic Media* (Durham, NC: Duke University Press, 2014).

22. Daston and Galison, *Objectivity*.

23. Steven Shapin, "Cordelia's Love: Credibility and the Social Studies of Science," *Perspectives on Science* 3, no. 3 (1995): 261.

24. Shapin, "Cordelia's Love," 262 (emphasis in original).

25. Thomas Kuhn, *The Structure of Scientific Revolutions*, 3rd ed. (Chicago: University of Chicago Press, 1996), 189.

26. Michael Polanyi, *The Tacit Dimension* (New York: Doubleday, 1966); Natalie Melas, *All the Difference in the World: Postcoloniality and the Ends of Comparison* (Palo Alto, CA: Stanford University Press, 2007).

27. Kuhn also introduced the term "disciplinary matrix" to explain how a network of social relationships could bind communities together. Thomas Kuhn, "Second Thoughts on Paradigms," in *The Essential Tension: Selected Studies in Scientific Tradition and Change,* ed. Frederick Suppe (Chicago: University of Chicago Press, 1977), 293–319. See also John Forrester, "If *p*, Then What? Thinking in Cases," *History of The Human Sciences* 9, no. 3 (1996): 1–25; Kuhn, *Structure of Scientific Revolutions*.

28. Michelle Murphy, *The Economization of Life* (Durham, NC: Duke University Press, 2017), 24.

29. Shannon Mattern, "Maintenance and Care," *Places Journal* (November 2018), https://placesjournal.org/article/maintenance-and-care/.

30. Gabrielle Hecht, "Interscalar Vehicles for an African Anthropocene: On Waste, Temporality, and Violence," *Cultural Anthropology* 33, no. 1 (2018): 109–141.

31. Lauren Berlant, "The Commons: Infrastructures for Troubling Times," *Environment and Planning D: Society and Space* 34, no. 3 (2016): 393–419.

32. Cait McKinney, *Information Activism: A Queer History of Lesbian Media Technologies* (Durham, NC: Duke University Press, 2020), 22.

33. Jacqueline Wernimont, *Numbered Lives: Life and Death in Quantum Media* (Cambridge, MA: MIT Press, 2019), 163.

34. Ari Luotonen and Kevin Altis, "World-Wide Web Proxies," *Computer Networks and ISDN Systems* 24, no. 2 (1994): 2 (emphasis added).

35. Markus Krajewski, *The Server*, trans. Ilinca Iurascu (New Haven, CT: Yale University Press, 2019); Fenwick McKelvey, *Internet Daemons: Digital Communications Possessed* (Minneapolis: University of Minnesota Press, 2018).

36. Susan Leigh Star and Karen Ruhleder, "Steps toward an Ecology of Infrastructure: Design and Access for Large Information Spaces," *Information Systems Research Information Systems Research* 7, no. 1 (1996): 111–134; Geoffrey C. Bowker, Karen Baker, Florence Miller, and David Ribes, "Toward Information Infrastructure Studies: Ways of Knowing in a Networked Environment," in *International Handbook of Internet Research*, eds. Jeremy Hunsinger, Lisbeth Klastrup, and Matthew M. Allen (Dordrecht, Netherlands: Springer, 2010), 97–117; Lisa Parks and Nicole Starosielski, eds., *Signal Traffic: Critical Studies of Media Infrastructures* (Urbana: University of Illinois Press, 2015); McKinney, *Information Activism*.

37. Thomas A. Stapleford, *The Cost of Living in America: A Political History of Economic Statistics, 1880–2000* (Cambridge: Cambridge University Press, 2010), 101.

38. Stapleford, *Cost of Living*.

39. In the mid-1990s, adjustments to the CPI took into account (in the case of cars) "improved corrosion protection, improved warranties, sealing improvements, stainless steel exhaust, longer-life spark plugs, improved steering gears, rust-resistant fuel injection, clearcoat paint, and more." Katharine G. Abraham, John S. Greenlees, and Brent R. Moulton, "Working to Improve the Consumer Price Index," *Journal of Economic Perspectives* 12, no. 1 (1998): 31.

40. Michael J. Boskin, Ellen L. Dulberger, Robert J. Gordon, Zvi Griliches, and Dale W. Jorgenson, "Consumer Prices, the Consumer Price Index, and the Cost of Living," *Journal of Economic Perspectives* 12, no. 1 (1998): 5.

41. Mayo Moran, "The Reasonable Person: A Conceptual Biography in Comparative Perspective," *Lewis & Clark Law Review*, 14 (2010): 1233.

42. The reasonable person's siblings come from John Gardner, "The Many Faces of the Reasonable Person," *Law Quarterly Review* 131 (2015): 563–584; "select group . . ." from *Helow v. Advocate General*, 1 WLR 2416 at 2417–2418 (2008).

43. Ellison Kahn, "A Trimestrial Potpourri," *South African Law Journal* 102, no. 1 (1985): 184–190.

44. Moran, "The Reasonable Person."

45. US Equal Employment Opportunity Commission on Harassment, https://www.eeoc.gov/laws/types/harassment.cfm (emphasis added).

46. Robert S. Lynd and Helen Merrell Lynd, *Middletown: A Study in American Culture* (New York: Harcourt, Brace, Jovanovich, 1956); *Middletown in Transition: A Study in Cultural Conflicts* (Boston: Houghton Mifflin Harcourt, 1982).

47. Sarah E. Igo, *The Averaged American: Surveys, Citizens, and the Making of a Mass Public* (Cambridge, MA: Harvard University Press, 2008), 55–56.

48. As Igo writes, in 1923, Lynd tried to explain his reasons for only including US-born whites in the Middletown studies: "The reason for this is obvious: since we are attempting a difficult new technique in a highly complicated field, it is desirable to simplify our situation as far as possible." *Averaged American*, 56.

49. Igo, *Averaged American*.

50. Elihu Katz and Paul F. Lazarsfeld, *Personal Influence: The Part Played by People in the Flow of Mass Communication* (Piscataway, NJ: Transaction Publishers, 1955), 335.

51. Katz and Lazarsfeld, *Personal Influence*, 339.

52. See Ensmenger, "Is Chess the Drosophila of Artificial Intelligence?" specifically page 6 and his comparison of artificial intelligence research to Robert Kohler's history of the use of *D. melanogaster* in genetics research.

53. Bowker et al., "Toward Information Infrastructure Studies."

54. I have mostly, here, focussed on entrenched proxies in technical, bureaucratic, and academic settings. But their presence is pervasive. While I discuss the artistic appropriation of proxies in chapter 6, Alice Christensen has reminded me of their appearance in literature, as well. Recall, for instance, the clerk (a "Prokurist"— a kind of legal proxy) in Franz Kafka's *Metamorphosis* who visits Gregor Samsa, and as proxy for his parents and boss, demands an explanation, "I am speaking here in the name of your parents and of your chief, and I beg you quite seriously to give me an immediate and precise explanation." Kafka chooses an actual legal proxy to stand in for the world *out there* at the threshold to Samsa's bedroom. Franz Kafka, *The Complete Stories*, ed. Nahum N. Glatzer (New York: Schocken, 1971), 97.

55. The term "fixed point" comes from the field of mathematics, but here, I am evoking the way that it is used to describe invariants in the process of developing a standard (a topic discussed at much greater length in chapter 2), in which fixed points become the hardened bases of a system of measurement and comparison. For a useful description of the scientific processes and debates behind setting fixed points, using the case of developing standardized thermometers, see Hasok Chang, *Inventing Temperature: Measurement and Scientific Progress* (Oxford: Oxford University Press, 2004).

56. Martha Lampland and Susan Leigh Star, *Standards and Their Stories: How Quantifying, Classifying, and Formalizing Practices Shape Everyday Life* (Ithaca, NY: Cornell University Press, 2009), 14.

57. From George William Francis, *The Dictionary of the Arts, Sciences, and Manufactures* (London: W. Brittain, 1846) (emphasis in original).

58. Stuart Hall, "New Ethnicities," in *Stuart Hall: Critical Dialogues in Cultural Studies*, eds. Kuan-Hsing Chen and David Morley (London: Routledge, 1996), 441–449.

59. Stanley Cavell, "The Uncanniness of the Ordinary," in *In Quest of the Ordinary: Lines of Skepticism and Romanticism* (Chicago: Chicago University Press, 1994), 172.

60. Berlant, "The Commons," 394.

61. Here, I am building on the work of others, including Mara Mills, Cait McKinney, Laine Nooney, and Jonathan Sterne, who offer ways of thinking of nonlinear media histories that do not efface the place of human bodies in technological cultures. See Mara Mills, "Do Signals Have Politics? Inscribing Abilities in Cochlear Implants." in *Oxford Handbook of Sound Studies*, ed. Trevor Pinch and Karin Bijsterveld (New York: Oxford University Press, 2011), 320–346; McKinney, *Information Activism*; Laine Nooney, "A Pedestal, a Table, a Love Letter: Archaeologies of Gender in Videogame History," *Game Studies* 13, no. 2 (2013), http://gamestudies.org/1302/articles/nooney; Jonathan Sterne, *MP3: The Meaning of a Format* (Durham, NC: Duke University Press, 2012).

62. In this way, proxies often function in a mediating role similar to quantum media, as described by Jacqueline Wernimont in *Numbered Lives: Life and Death in Quantum Media* (Cambridge, MA: MIT Press: 2019).

63. John Durham Peters, "Witnessing," *Media, Culture & Society* 23, no. 6 (2001): 709; Carrie Rentschler, "Witnessing: US Citizenship and the Vicarious Experience of Suffering" *Media, Culture & Society* 26, no. 2 (2004): 296–304.

64. This quote is from the June 18, 1999, issue of the paper. The article "Bombs Away at Yodaville," written by James W. Crawley, was digitized and uploaded to a Geocities site with the improbable URL of www.geocities.com/pentagon, which at the time was the home page containing information and resources on military operations on urbanized terrain. If you want to read the whole article, you might be able to view it here: https://web.archive.org/web/20020207204247/http://www.geocities.com:80/Pentagon/6453/index.html.

65. On the construction of testing scenarios, see Donald Mackenzie, *Inventing Accuracy* (Cambridge, MA: MIT Press, 1990); Trevor Pinch, "'Testing—One, Two, Three . . . Testing!': Toward a Sociology of Testing," *Science, Technology, & Human Values* 18, no. 1 (1993): 25–41.

66. Glenn et al., *Preparing for the Proven Inevitable*; Christine Hoekenga, "3:10 to Baghdad," *High Country News* (March 31, 2008), https://www.hcn.org/issues/367/17605. These place names are also repeated in Microsoft PowerPoint slides from the US Military.

67. On immersion, military training, and simulation, see Lucy Suchman, "Configuring the Other: Sensing War through Immersive Simulation," *Catalyst: Feminism, Theory, Technoscience* 2, no. 1 (2016): 1–36.

68. "Iraq War Ratchets up Work at Yuma-Area Bases," *Tucson Citizen*, March 23, 2006.

69. Pinch, "Testing," 26.

70. It is worth mentioning that Yuma, Arizona, and the surrounding desert has also been a frequent filming location, since the earliest days of the American film industry. The desert has stood in for other terrestrial deserts as well as the desert planet Tatooine in the Star Wars film, *Return of the Jedi*.

71. Glenn et al., *Preparing for the Proven Inevitable*, 38–39; Susan Leigh Star, "The Ethnography of Infrastructure," *American Behavioral Scientist* 43, no. 3 (1999): 377–391.

72. Richard Schechner, *Between Theatre and Anthropology* (Philadelphia: University of Philadelphia Press, 1985); Richard Schechner, *Performance Studies: An Introduction* (New York: Routledge, 2002); Rebecca Schneider, *Performing Remains: Art and War in Times of Theatrical Reenactment* (London and New York: Taylor & Francis, 2011).

73. Vaihinger, *The Philosophy of "As If."*

74. Vaihinger provides this example of fictions as arbitrary-but-necessary reference points: "Here we may also include all the arbitrary determinations found in science, such as, for example, the meridian of Ferro, the determination of the zero point, the selection of water as the measure of specific gravity, of the movements of the stars as an index of time. In all these cases certain points of reference are taken and lines similar to co-ordinates drawn in different directions for the determination and classification of phenomena." *The Philosophy of "As If,"* 23–24.

75. Vaihinger, *The Philosophy of "As If,"* 15 (emphasis added).

76. Anthony Appiah, *As If: Idealization and Ideals* (Cambridge, MA: Harvard University Press, 2017), 3.

77. Vaihinger argues that fictitious means predominate in disciplines that must manage "a large number of quantities that oscillate around an ideal point (e.g., meteorology and statistics)," *The Philosophy of "As If,"* 224, n2.

78. William Stanley Jevons, *Principles of Science: A Treatise on Logic and Scientific Method* (New York: MacMillan, 1874), 422.

79. Theodore Porter, *The Rise of Statistical Thinking, 1820–1900* (Princeton, NJ: Princeton University Press, 1988), 52; Lambert Adolphe Jacques Quetelet, *A Treatise on Man and the Development of His Faculties*, ed. T. Smibert, trans. R. Knox (Cambridge: Cambridge University Press, 2013).

80. Georges Canguilhem, *The Normal and the Pathological*, trans. Carolyn R. Fawcett (Brooklyn: Zone Books, 1989); Lennard J. Davis, *Enforcing Normalcy: Disability, Deafness, and the Body* (London and New York: Verso, 1995); Michael Warner, *The Trouble with Normal: Sex, Politics, and the Ethics of Queer Life* (New York: Free Press, 1999).

81. Ian Hacking, *The Taming of Chance* (Cambridge: Cambridge University Press, 1990).

82. Canguilhem, *The Normal and the Pathological*; Hacking, *The Taming of Chance.*

83. See, for instance Aimi Hamraie's history of design thinking, and the discussion of Henry Dreyfuss's user designs based on statistical averages. Aimi Hamraie, *Building Access: Universal Design and the Politics of Disability* (Minneapolis: University of Minnesota Press, 2017); Henry Dreyfuss, *The Measure of Man: Human Factors in Design*, 2nd ed. (New York: Whitney Library of Design, 1967).

84. Hacking, *The Taming of Chance.*

85. Hacking, *The Taming of Chance*, 108.

86. Davis, *Enforcing Normalcy*; Hamraie, *Building Access*.

87. Lawrence Busch, *Standards: Recipes for Reality* (Cambridge, MA: MIT Press, 2011).

88. On the idea that things that are felt to be real are real in their effects, see the Thomas Theorem, first articulated by William I. Thomas and Dorothy S. Thomas in *The Child in America: Behavior Problems and Programs* (New York: Alfred A. Knopf, 1928); see also Geoffrey C. Bowker and Susan Leigh Star, *Sorting Things Out: Classification and Its Consequences* (Cambridge, MA: MIT Press, 1999). Vaihinger was aware of the similarity of his theory to pragmatism, writing:

   Pragmatism, too, so widespread throughout the English-speaking world, has done something to prepare the ground for Fictionalism, in spite of their fundamental difference. Fictionalism does not admit the principle of Pragmatism which runs: 'An idea which is found to be useful in practice proves thereby that it is also true in theory, and the fruitful is thus always true'. The principle of Fictionalism, on the other hand, or rather the outcome of Fictionalism, is as follows: An idea whose theoretical untruth or incorrectness, and therewith its falsity, is admitted, is not for that reason practically valueless and useless; for such an idea, in spite of its theoretical nullity may have great practical importance. (*The Philosophy of "As If,"* viii)

89. Bose et al. showed in 2011 that injury and fatality rates between people categorized as male versus female showed that "the odds for a belt-restrained female driver to sustain severe injuries were 47% higher than those for a belt-restrained male driver involved in a comparable crash." Dipan Bose, Maria Segui-Gomez, and Jeff R. Crandall, "Vulnerability of Female Drivers Involved in Motor Vehicle Crashes: An Analysis of US Population at Risk," *American Journal of Public Health* 101 (December 2011): 2368–2373. See also Caroline Criado-Perez, *Invisible Women: Exposing Data Bias in a World Designed for Men* (New York: Harry N. Abrams, 2019).

90. While the goal of Bose et al. was to demonstrate sex-specific disparities in injury prevention, we should resist the urge to sort people along dichotomous, sex-specific categories, and instead see the body-specific ways that a reliance on an anthropomorphic model based on so-called averageness has injurious outcomes for many people. Bose et al., "Vulnerability of Female Drivers."

91. Bowker and Star, *Sorting Things Out*; John Dewey, *The Essential Dewey*, ed. Larry A. Hickman and Thomas M. Alexander (Bloomington: Indiana University Press, 1998).

92. Bowker and Star, *Sorting Things Out*, 290.

93. Hamraie, *Building Access*; Rosemary Garland Thomson, *Extraordinary Bodies: Figuring Physical Disability in American Culture and Literature* (New York: Columbia University Press, 2017).

94. Judith Butler, "Performative Acts and Gender Constitution: An Essay in Phenomenology and Feminist Theory," *Theatre Journal* 40, no. 4 (1988): 519–531.

95. Friedrich Nietzsche, *The Portable Nietzsche*, trans. Walter Kaufmann (New York: Penguin Books, 1982), 47.

96. Charles Goodwin, "Professional Vision," *American Anthropologist* 96, no. 3 (1994): 606–633.

97. For some recent, like-minded work on gendering as a process within technological cultures, see Wendy Chun, *Programmed Visions: Software and Memory* (Cambridge, MA: MIT Press, 2011); Jennifer S. Light, "When Computers Were Women," *Technology and Culture* 40, no. 3 (1999): 455–483; Amy Adele Hasinoff, *Sexting Panic: Rethinking Criminalization, Privacy, and Consent* (Urbana: University of Illinois Press, 2015); Lisa Nakamura, "Indigenous Circuits: Navajo Women and the Racialization of Early Electronic Manufacture," *American Quarterly* 66, no. 4 (2014): 919–941; Rena Bivens, "The Gender Binary Will Not Be Deprogrammed: Ten Years of Coding Gender on Facebook," *New Media & Society* 9, no. 6 (2017): 880–898; Mar Hicks, *Programmed Inequality: How Britain Discarded Women Technologists and Lost Its Edge in Computing* (Cambridge, MA: MIT Press, 2017).

98. Michel Foucault, *The Order of Things: An Archaeology of the Human Sciences*, trans. A. M. Sheridan (London: Routledge, 2002[1970]).

CHAPTER 2

1. Carolyn Marvin, *When Old Technologies Were New: Thinking about Electric Communication in the Late Nineteenth Century* (Oxford: Oxford University Press, 1988).

2. The NPL is the national measurement institute in the United Kingdom and a crucial part of the country's national measurement infrastructure.

3. Michael de Podesta, "The Measure of Science: Redefining the Kilogram," presentation, the Royal Institution, London, October 22, 2018.

4. This particular kilogram lasted 130 years, but the idea of a physical mass standard in the metric system persisted for almost 220 years.

5. de Podesta, "The Measure of Science."

6. de Podesta, "The Measure of Science."

7. de Podesta, "The Measure of Science."

8. Although the Planck constant did not previously have a fixed value (only a value that included a standard uncertainty), the new instructions for "realizing" kilograms required three sufficiently precise measurements (5 parts in 10^8) of the Planck constant using a watt balance. Once the three measurements were obtained, they were averaged out to create a new, fixed value of the Planck constant, with no standard uncertainty. Philippe Richard, Hao Fang, and Richard Davis, "Foundation for the Redefinition of the Kilogram," *Metrologia* 53, no. 5 (2016): A6.

9. Terry Quinn, *From Artefacts to Atoms: The BIPM and the Search for Ultimate Measurement Standards* (Oxford and New York: Oxford University Press, 2011), 341.

10. David Turnbull, "The Ad Hoc Collective Work of Building Gothic Cathedrals with Templates, String, and Geometry," *Science, Technology, & Human Values* 18, no. 3 (1993): 315–340.

11. Émile Durkheim, *The Elementary Forms of Religious Life*, trans. Karen E. Fields (New York: Free Press, 1995[1912]).

12. Lawrence Busch, *Standards: Recipes for Reality* (Cambridge, MA: MIT Press, 2011). See also Martha Lampland and Susan Leigh Star, eds., *Standards and Their Stories: How Quantifying, Classifying, and Formalizing Practices Shape Everyday Life* (Ithaca, NY: Cornell University Press, 2009).

13. Ludwig Wittgenstein, *Philosophical Investigations*, part 1, §108. Many thanks to Nick Couldry for pointing me to this passage (the emphasis is added).

14. Mary Douglas, *Purity and Danger: An Analysis of Concepts of Pollution and Taboo* (New York: Routledge, 2002).

15. Douglas, *Purity and Danger*, 2.

16. Lisa Gitelman and Virginia Jackson, "Introduction," in *"Raw Data" Is an Oxymoron*, ed. Lisa Gitelman (Cambridge, MA: MIT Press, 2013), 7.

17. In their account of Irving Fisher's attempts to represent economic models, Mary Poovey and Kevin Brine excavate the laboriousness of data cleaning. Mary Poovey and Kevin Brine, "From Measuring Desire to Quantifying Expectations: A Late Nineteenth-Century Effort to Marry Economic Theory and Data," in Gitelman, *"Raw Data" Is an Oxymoron*, 61–88.

18. Jean-Christophe Plantin, "Data Cleaners for Pristine Datasets: Visibility and Invisibility of Data Processors in Social Science," *Science, Technology, & Human Values* 44, no. 1 (2019): 52–73.

19. Douglas, *Purity and Danger*, 2.

20. Bob Woodward and Carl Bernstein, "Funds Laundered," *Washington Post*, July 11, 1974.

21. Bob Woodward and Carl Bernstein, *All the President's Men* (New York: Simon & Schuster, 1974), 46–47.

22. The *Oxford English Dictionary*'s definition of "(money) laundering" is as follows: "To transfer funds of dubious or illegal origin, usu. to a foreign country, and then later to recover them from what seem to be 'clean' (i.e. legitimate) sources. The use arose from the Watergate inquiry in the United States in 1973–4." It should be noted that though etymologies of "money laundering" identify the Watergate scandal as the first appearance, my research shows that the term was not only already in circulation before the scandal, it actually appeared in print in reference to the Mafia before it appeared in reference to Watergate. See James M. Markham, "Mob-Influenced Businesses Would Fill a List from A to Z, Officials Here Say," *New York Times*, August 19, 1972.

23. Lana Swartz, *New Money: How Payment Became Social Media* (New Haven, CT: Yale University Press, 2020).

24. Geoffrey C. Bowker, *Memory Practices in the Sciences* (Cambridge, MA: MIT Press, 2005).

25. Douglas, *Purity and Danger*, 2.

26. Matt Ford, "Use It or Lose It?" *Atlantic Monthly*, May 30, 2017.

27. J. Christian Adams, *A Primer on "Motor Voter": Corrupted Voter Rolls and the Justice Department's Selective Failure to Enforce Federal Mandates.* Heritage Foundation, September 25, 2014.

28. Adams, *A Primer on "Motor Voter"* (emphasis added).

29. Michelle Murphy, *Sick Building Syndrome and the Problem of Uncertainty* (Durham, NC: Duke University Press, 2006), 19.

30. Craig Robertson, *The Filing Cabinet: A Vertical History of Information* (Minneapolis: University of Minnesota Press, 2021); Craig Robertson, "Learning to File: Reconfiguring Information and Information Work in the Early Twentieth Century," *Technology and Culture* 58, no. 4 (2017): 955–981.

31. Robertson, *The Filing Cabinet*, 51.

32. Ruth Schwartz Cowan, *More Work for Mother* (New York: Basic Books, 1983).

33. Stephen Prince, *Classical Film Violence: Designing and Regulating Brutality in Hollywood Cinema, 1930–1968* (New Brunswick, NJ: Rutgers University Press, 2003); Tarleton Gillespie, *Custodians of the Internet: Platforms, Content Moderation, and the Hidden Decisions That Shape Social Media* (New Haven, CT: Yale University Press, 2018); Sarah T. Roberts, *Behind the Screen* (New Haven, CT: Yale University Press, 2019).

34. Raiford Guins, *Edited Clean Version: Technology and the Culture of Control* (Minneapolis: University of Minnesota Press, 2008).

35. Rena Bivens, "The Gender Binary Will Not Be Deprogrammed: Ten Years of Coding Gender on Facebook," *New Media & Society* 19, no. 6 (2017): 880–898.

36. On content moderation and user experience, see Ysabel Gerrard and Helen Thornham, "Content Moderation: Social Media's Sexist Assemblages," *New Media & Society* 22, no. 7 (2020): 1266–1286; Gillespie, *Custodians of the Internet*; Roberts, *Behind the Screen*; Sarah Myers West, "Censored, Suspended, Shadowbanned: User Interpretations of Content Moderation on Social Media Platforms," *New Media & Society* 20, no. 11 (2018): 4366–4383. On the policing of sex work online, see Kendra Albert et al., *FOSTA in Legal Context* (July 30, 2020), https://papers.ssrn.com/sol3/papers.cfm?abstract_id=3663898.

37. T. L. Cowan, "Digital Hygiene: A Metaphor of Dirty Proportions," http://www.drecollab.org/digital-hygiene-a-metaphor-of-dirty-proportions/.

38. Daniela Agostinho and Nanna Bonde Thylstrup, "'If Truth Was a Woman': Leaky Infrastructures and the Gender Politics of Truth-Telling," *Ephemera* 19, no. 4 (2019): 745–775.

39. Thomas P. Hughes, "The evolution of large technological systems," in *The Social Construction of Technological Systems: New Directions in the Sociology and History of Technology*, eds.

Wiebe E. Bijker, Thomas P. Hughes, and Trevor Pinch (Cambridge, MA: MIT Press, 1987), 51–82.

40. For a tiny slice of this literature, see Geoffrey C. Bowker and Susan Leigh Star, *Sorting Things Out: Classification and Its Consequences* (Cambridge, MA: MIT Press, 1999); Busch, *Standards*; Laura DeNardis, *Protocol Politics: The Globalization of Internet Governance* (Cambridge, MA: MIT Press, 2009); Laura DeNardis, *Opening Standards: The Global Politics of Interoperability* (Cambridge, MA: MIT Press, 2011); Alexander R. Galloway, *Protocol: How Control Exists after Decentralization* (Cambridge, MA: MIT Press, 2004); Thomas P. Hughes, *Networks of Power: Electrification in Western Society, 1880–1930* (Baltimore: Johns Hopkins University Press, 1983); Lampland and Star, *Standards and Their Stories*; Thomas S. Mullaney, "The Moveable Typewriter: How Chinese Typists Developed Predictive Text during the Height of Maoism," *Technology and Culture* 53, no. 4 (2012): 777–814; David Noble, *America by Design* (New York: Knopf, 1982); Andrew Russell, *Open Standards and the Digital Age: History, Ideology, and Networks* (Cambridge: Cambridge University Press, 2014); Amy Slaton and Janet Abbate, "The Hidden Lives of Standards: Technical Prescriptions and the Transformation of Work in America," in *Technologies of Power* (Cambridge, MA: MIT Press, 2001): 95–144; Susan Leigh Star, "The Ethnography of Infrastructure," *American Behavioral Scientist* 43, no. 3 (1999): 377–391; Nicole Starosielski, *The Undersea Network* (Durham, NC: Duke University Press, 2015); Jonathan Sterne, *MP3: The Meaning of a Format* (Durham, NC: Duke University Press, 2012); JoAnne Yates and Craig N. Murphy, *Engineering Rules: Global Standard Setting since 1880* (Baltimore: Johns Hopkins University Press, 2019).

41. Russell, *Open Standards*.

42. Elizabeth Cullen Dunn, "Standards without Infrastructure," in *Standards and Their Stories: How Quantifying, Classifying, and Formalizing Practices Shape Everyday Life,* eds. Martha Lampland and Susan Leigh Star (Ithaca, NY: Cornell University Press, 2009), 119.

43. Hasok Chang, *Inventing Temperature: Measurement and Scientific Progress* (New York: Oxford University Press, 2004).

44. For more on the scientific-historical dimensions of precision (and, for instance, its contrast with accuracy), see M. Norton Wise, ed., *The Values of Precision* (Princeton, NJ: Princeton University Press, 1997); Busch, *Standards*; Ken Alder, "Making Things the Same," *Social Studies of Science* 28, no. 4 (1998): 499–545.

45. This particular conundrum comes from a discussion with Jonathan Sterne.

46. Hans Vaihinger, *The Philosophy of "As If,"* trans. C. K. Ogden (New York: Harcourt Brace, 1924).

47. Busch, *Standards*.

48. Busch, *Standards*.

49. As DeNardis writes about interoperability: "Think of the 19th-century development of a standard gauge interconnecting railroads across the vast expanse of North America or, more

than a century later, the standardization that allows us to exchange e-mail across different types of applications and hardware devices." DeNardis, "The Social Stakes of Interoperability," *Science* 337, no. 6101 (2012): 1454.

50. Tarleton Gillespie, *Wired Shut: Copyright and the Shape of Digital Culture* (Cambridge, MA: MIT Press, 2007); DeNardis, *Opening Standards*.

51. Dean Spade, *Normal Life: Administrative Violence, Critical Trans Politics, and the Limits of Law* (Durham, NC: Duke University Press, 2011).

52. Lampland and Star, *Standards and Their Stories.*

53. Lampland and Star, *Standards and Their Stories.*

54. Bowker and Star, *Sorting Things Out.*

55. Nancy Fraser, *Justice Interruptus: Critical Reflections on the "Postsocialist" Condition* (New York: Routledge, 1997).

56. Geof Bowker, "How to Be Universal: Some Cybernetic Strategies, 1943–70," *Social Studies of Science* 23, no. 1 (1993): 107–127.

57. Kathryn M. Olesko, "The Meaning of Precision: The Exact Sensibility in Early-Nineteenth-Century Germany," in *The Values of Precision,* ed. M. N. Wise (Princeton, NJ: Princeton University Press, 1995): 103.

58. James C. Scott, *Seeing Like a State: How Certain Schemes to Improve the Human Condition Have Failed* (New Haven, CT: Yale University Press, 1998), 4.

59. See Olesko, "The Meaning of Precision," in particular 117–118.

60. Sandford Fleming, *Time-Reckoning for the Twentieth Century* (Washington, DC: Smithsonian, 1889); Stephen Kern, *The Culture of Time and Space, 1880–1918* (Cambridge, MA: Harvard University Press, 1983). For more discussion of fixed units based on the measurement of time and the utopian dreams of such a standard, see Dylan Mulvin, "The Media of High-Resolution Time: Temporal Frequencies as Infrastructural Resources," *The Information Society* 33, no. 5 (2017): 282–290.

61. Max Planck, "Über Irreversible Strahlungsvorgänge" [On Irreversible Radiation Processes], *Annalen der Physik* 306, no. 1 (1900): 121.

62. These instructions, including the supplies, estimated times, and figures, are based on G. Girard, "The Washing and Cleaning of Kilogram Prototypes at the BIPM," *BIPM Internal Report* (1990). "Rather handsome, but not specular" is from Richard Davis, "Recalibration of the U.S. National Prototype Kilogram," *Journal of Research of the National Bureau of Standards* 90, no. 4 (1985): 265. I have added certain stylistic emphases to this text.

63. Quinn, *From Artefacts to Atoms.*

64. Specifically, "one of these is kept by the Director of the BIPM, one is in the possession of the President of the CIPM and the third is held by the Archives de France." Richard Davis, "The SI Unit of Mass," *Metrologia* 40, no. 6 (2003): 300.

65. In the 1980s, three keys were required to launch a Trident (nuclear) missile from a US Navy submarine. But Hollywood settled on two keys long ago. See, for instance, *The Hunt for Red October,* dir. John McTiernan (Los Angeles: Paramount, 1990). See also Gerald Marsh, "Skirting Human Error: The Navy's Missile Launch System," *Bulletin of Atomic Scientists* 43, no. 1 (1987): 38.

66. Stuart Davidson, "A Review of Surface Contamination and The Stability of Standard Masses," *Metrologia* 40, no. 6 (2003): 324–338.

67. Ken Alder, *The Measure of All Things: The Seven-Year Odyssey and Hidden Error That Transformed the World* (New York: Free Press, 2002); Ken Alder, "Revolution to Measure: The Political Economy of the Metric System in France," in *The Values of Precision*, ed. M. Norton Wise (Princeton, NJ: Princeton University Press, 1995): 39–71.

68. Joint Committee for Guides in Metrology (JCGM), *International Vocabulary of Metrology,* 3rd ed. (2008), 241, https://www.bipm.org/utils/common/documents/jcgm/JCGM_200 _2012.pdf.

69. David Freedman, Robert Pisani, and Roger Purves, *Statistics* (New York: W. W. Norton, 2007), 98.

70. National laboratories, which are government agencies responsible for maintaining scientific, technical, and sometimes consumer standards, are responsible for checking their standards against those kept by international bodies. They also disseminate mass standards to lower positions on the hierarchy.

71. Maurice Crosland, "The Congress on Definitive Metric Standards, 1798–1799: The First International Scientific Conference?" *Isis* 60, no. 2 (1969): 226–231.

72. Alder, *The Measure of All Things.*

73. Alder, *The Measure of All Things,* 21 (emphasis added).

74. Alder, *The Measure of All Things,* 19; Alder, "Revolution to Measure," 41.

75. Alder, *The Measure of All Things,* 20.

76. Scott, *Seeing Like a State,* 30.

77. Robert P. Crease, *World in the Balance: The Historic Quest for An Absolute System of Measurement* (New York: W. W. Norton, 2012).

78. Quinn, *From Artefacts to Atoms.*

79. Johnson & Matthey was the assayer and refiner of the Bank of England. See Davis, "Recalibration."

80. Quinn, *From Artefacts to Atoms.*

81. From the BIPM's frequently asked questions about the international prototype of the kilogram, http://www.bipm.org/en/bipm/mass/faqs_mass.html.

82. Peter Galison, *Einstein's Clocks and Poincaré's Maps: Empires of Time* (New York: W. W. Norton, 2003).

83. *Comptes Rendus des Séances de la Première Conférence Générale des Poids et Mesures, Réunie a Paris en 1889* (Paris: Guathier-Villars et Fils, Imprimeures-Libraires, 1890); Craig Robertson, *The Passport in America: The History of a Document* (New York: Oxford University Press, 2010).

84. There was some debate over which French governmental office was worthy of having custody of the key. *Comptes Rendus.*

85. Jacques Derrida, "Declarations of Independence," *New Political Science* 7, no. 1 (1986): 10.

86. Bureau International des Poids et Mesures (BIPM), *SI Brochure: The International System of Units (SI)*, 8th ed. (Paris: Stedi Media, 2006), 112.

87. Quinn, *From Artefacts to Atoms.*

88. Georges Canguilhem notes that in France in the 1930s, "normalization" was used in place of "standardization," as if the two had always been interchangeable. The IPK puts this elision into practice by making all nonprototype kilograms inherent deviations. *The Normal and the Pathological,* trans. Carolyn R. Fawcett (Brooklyn: Zone Books, 1989), 247.

89. Prior to its redefinition in 1960, the definition of length was based on the Prototype Meter. It is now based on the length of light traveling in a vacuum in a certain fraction of a second.

90. Ludwig Wittgenstein, *Philosophical Investigations* (Oxford, UK: B. Blackwell, 1953), §50.

91. Nathan Salmon, "How to Measure the Standard Metre," *Proceedings of the Aristotelian Society* 88 (1987): 193–217. Charles Peirce also turned to physical standards in explicating his science of signs: "A yard-stick might seem, at first sight, to be an icon of a yard; and so it would be, if it were merely intended to show a yard as near as it can be seen and estimated to be a yard. But the very purpose of a yard-stick is to show a yard nearer than it can be estimated by its appearance. This it does in consequence of an accurate mechanical comparison made with the bar in London called the yard. Thus it is a real connection which gives the yard-stick its value as a representamen; and thus it is an *index*, not a mere *icon*." *Philosophical Writings of Peirce*, ed. Justus Buchler (New York: Dover Publications, 1955), 109 (emphasis in original).

92. Natalie Melas, *All the Difference in the World: Postcoloniality and the Ends of Comparison* (Stanford, CA: Stanford University Press, 2007).

93. Karen Barad, *Meeting the Universe Halfway: Quantum Physics and the Entanglement of Matter and Meaning* (Durham, NC: Duke University Press, 2007), 19 (emphasis in original).

94. Barad, *Meeting the Universe Halfway*, 19.

95. Comité International des Poids et Mesures (CIPM), *Procès-Verbaux des Séances de 1882* (Paris: Gauthier-Villars, 1883); Quinn, *From Artefacts to Atoms.*

96. Quinn, *From Artefacts to Atoms.*

97. Girard, "The Washing and Cleaning of Kilogram Prototypes"; Quinn, *From Artefacts to Atoms.*

98. Davis, "The SI Unit of Mass," 303.

99. BIPM, *SI Brochure*, 112.

100. Peter Cumpson and Sano Naoko, "Stability of Reference Masses V: UV/Ozone Treatment of Gold and Platinum Surfaces," *Metrologia* 50, no. 1 (2013): 27–36; P. J. Cumpson and M. P. Seah, "Stability of Reference Masses I: Evidence for Possible Variations in the Mass of Reference Kilograms Arising from Mercury Contamination," *Metrologia* 31, no. 1 (1994): 21–26.

101. Andrew Barry, "Materialist Politics: Metallurgy," in *Political Matter: Technoscience, Democracy, and Public Life*, eds. Bruce Braun and Sarah Whatmore (Minneapolis: University of Minnesota Press, 2010), 90.

102. Crease, *World in the Balance*, 207.

103. Steven Jackson and David Ribes, "Data Bit Man: The Work of Sustaining a Long-Term Study," in *"Raw Data" Is an Oxymoron*, 163.

104. Quinn, *From Artefacts to Atoms*, 173.

105. Bruno Latour, "Mixing Humans and Nonhumans Together: The Sociology of a Door-Closer," *Social Problems* 35, no. 3 (1988): 300.

106. Thomas Gieryn, "What Buildings Do," *Theory and Society* 31, no.1 (2002): 35–74.

107. Trevor Pinch, "'Testing—One, Two, Three . . . Testing!': Toward a Sociology of Testing," *Science, Technology, & Human Values* 18, no. 1 (1993): 25–26.

CHAPTER 3

1. See "The Lenna Story," http://www.lenna.org/.

2. This is the story as told to Peter Nowak in *Sex, Bombs, and Burgers: How War, Pornography, and Fast Food Have Shaped Modern Technology* (Guilford, CT: Rowman & Littlefield, 2011), 173 (emphasis added).

3. Jamie Hutchinson, "Culture, Communication, and an Information Age Madonna," *IEEE Professional Communication Society Newsletter* 45, no. 3 (2001): 1 (emphasis added).

4. Susanna Paasonen, Kylie Jarrett, and Ben Light, *NSFW: Sex, Humor, and Risk in Social Media* (Cambridge, MA: MIT Press, 2019).

5. Eve Kosofsky Sedgwick, *Between Men: English Literature and Male Homosocial Desire* (New York: Columbia University Press, 1985).

6. Lauren Berlant, *The Queen of America Goes to Washington City: Essays on Sex and Citizenship* (Durham, NC: Duke University Press, 1997), 59.

7. Here, I am adapting Simidele Dosekun's "spectacular femininity" to highlight the conspicuous ways that heterosexual masculinity is performed and transformed into a spectacle for the consumption of other men. Simidele Dosekun, *Fashioning Postfeminism: Spectacular Femininity and Transnational Culture* (Urbana: University of Illinois Press, 2020).

8. Nathan Ensmenger, *The Computer Boys Take Over* (Cambridge, MA: MIT Press, 2010); Nathan Ensmenger, "'Beards, Sandals, and Other Signs of Rugged Individualism': Masculine Culture within the Computing Professions," *Osiris* 30, no. 1 (2015): 38–65.

9. Jacqueline Wernimont, *Numbered Lives: Life and Death in Quantum Media* (Cambridge, MA: MIT Press, 2019); Paul Dourish, *The Stuff of Bits: An Essay on the Materialities of Information* (Cambridge, MA: MIT Press, 2017); Matthew Kirschenbaum, *Mechanisms: New Media and the Forensic Imagination* (Cambridge, MA: MIT Press, 2008).

10. Lisa Gitelman and Virginia Jackson, "Introduction," in *"Raw Data" Is an Oxymoron*, ed. Lisa Gitelman (Cambridge, MA: MIT Press, 2013), 3.

11. Kathryn Henderson, "The Visual Culture of Engineers," *Sociological Review* 42, no. 1 (1994): 196–218.

12. Li Cornfeld, "Babes in Tech Land: Expo Labor as Capitalist Technology's Erotic Body" *Feminist Media Studies* 18, no. 2 (2018): 205–220; Cait McKinney, *Information Activism: A Queer History of Lesbian Media Technologies* (Durham, NC: Duke University Press, 2020); Amy Adele Hasinoff, *Sexting Panic: Rethinking Criminalization, Privacy, and Consent* (Urbana: University of Illinois Press, 2015); Sharif Mowlabocus, *Gaydar Culture* (London: Ashgate/Routledge, 2010); Kate O'Riordan and David J Phillips, eds., *Queer Online: Media Technology and Sexuality* (Bern, Switzerland: Peter Lang, 2007).

13. Hortense J. Spillers, "Mama's Baby, Papa's Maybe: An American Grammar Book," *Diacritics* 17, no. 2 (1987): 67; Wendy Hui Kyong Chun, *Control and Freedom: Power and Paranoia in the Age of Fiber Optics* (Cambridge, MA: MIT Press, 2006); Safiya Umoja Noble, *Algorithms of Oppression* (New York: NYU Press, 2018); Lisa Nakamura, *Digitizing Race: Visual Cultures of the Internet* (Minneapolis: University of Minnesota Press, 2008).

14. On the materiality of gendered power and its intersection with professionalization, see Cynthia Cockburn, "The Material of Male Power," *Feminist Review* 9, no. 1 (1981): 41–58.

15. Sara Ahmed, *Queer Phenomenology: Orientations, Objects, Others* (Durham, NC: Duke University Press, 2006), 40 (emphasis added).

16. Paul Du Gay, Stuart Hall, Linda Janes, Anders Koed Madsen, Hugh Mackay, and Keith Negus, *Doing Cultural Studies: The Story of the Sony Walkman* (London: SAGE, 2013); Lorraine Daston, ed., *Things That Talk* (Brooklyn: Zone Books, 2004).

17. Marilyn Strathern, *Reproducing the Future: Anthropology, Kinship, and the New Reproductive Technologies* (London: Routledge, 1992), 33.

18. Charles Goodwin, "Professional Vision," *American Anthropologist* 96, no. 3 (1994): 606.

19. Goodwin, "Professional Vision," 626.

20. Goodwin's work has traveled far and wide in the study of the embodied practices of professional vision. See, for instance, the work of Janet Vertesi, who documents the construction of professional vision within the Mars Rover program. She clarifies that "professional

vision" is not only a matter of what one does with one's eyes; rather, it is a range of inter-related practices through which meaning is created through learned skill and embodied technique. *Seeing Like a Rover: How Robots, Teams, and Images Craft Knowledge of Mars* (Chicago: University of Chicago Press, 2015).

21. Jacob Gaboury, "Image Objects: An Archaeology of 3D Computer Graphics, 1965–1979," PhD dissertation, New York University, 2015; Ann-Sophie Lehmann, "Taking the Lid off the Utah Teapot towards a Material Analysis of Computer Graphics," *Zeitschrift für Medien- und Kulturforschung*, no. 1 (2012): 169–184; Michael Baxandall, *Painting and Experience in Fifteenth-Century Italy: A Primer in the Social History of Pictorial Style* (Oxford: Oxford University Press, 1988).

22. Here is an elegant example of how this happens: The Lena image has two Wikipedia pages, one for the centerfold and one for the test image. Depending on the day and the backstage debates among editors, the image will sometimes appear on the test image page (because it is fair use) but not on the centerfold page. See https://en.wikipedia.org/wiki/Lenna.

23. Jonathan Sterne *MP3: The Meaning of a Format* (Durham, NC: Duke University Press, 2012); Dourish, *The Stuff of Bits*.

24. Dylan Mulvin and Jonathan Sterne, "Scenes from an Imaginary Country: Test Images and the American Color Television Standard," *Television & New Media* 17, no. 1 (2016): 21–43.

25. David Salomon, *Data Compression: The Complete Reference* (London: Springer, 2007), 517.

26. Shea Swauger, "Software That Monitors Students during tests Perpetuates Inequality and Violates Their Privacy," *MIT Technology Review* (August 7, 2020), https://www.technologyreview.com/2020/08/07/1006132/software-algorithms-proctoring-online-tests-ai-ethics/.

27. The experience, and explanation for this conflict, is described at-length by Sasha Constanza-Chock in a description of "traveling while trans." Sasha Costanza-Chock, *Design Justice: Community-Led Practices to Build the Worlds We Need* (Cambridge, MA: MIT Press, 2020).

28. David G. Lowe "Object Recognition from Local Scale-Invariant Features," *Proceedings of the Seventh IEEE International Conference on Computer Vision*, 2 (1999): 1150–1157.

29. Nick Seaver, "Algorithms as Culture: Some Tactics for the Ethnography of Algorithmic Systems," Big Data & Society 4, no. 2 (2017): https://doi.org/10.1177/2053951717738104.

30. Ruha Benjamin, *Race after Technology: Abolitionist Tools for the New Jim Code* (Cambridge, UK: Polity, 2019).

31. Simone Browne, *Dark Matters* (Durham, NC: Duke University Press, 2015); Benjamin, *Race after Technology*; Joy Buolamwini and Timnit Gebru, "Gender Shades: Intersectional Accuracy Disparities in Commercial Gender Classification," *Conference on Fairness, Accountability and Transparency*, PMLR 81 (2018): 77–91. In addition, see the work of the Algorithmic Justice League (https://www.ajl.org).

32. Browne, *Dark Matters*.

33. Stuart Hall, "The Whites of Their Eyes: Racist Ideologies and the Media," in *Gender, Race, and Class in Media*, eds. Gail Dines and Jean M. Humez (Thousand Oaks, CA: SAGE, 1995), 18–22; Meredith Broussard, *Artificial Unintelligence: How Computers Misunderstand the World* (Cambridge, MA: MIT Press, 2018); Nakamura, *Digitizing Race*.

34. Lewis Gordon, "Is the Human a Teleological Suspension of Man?" in *After Man, towards the Human: Critical Essays on Sylvia Wynter*, ed. Anthony Bogues (Kingston, Jamaica: Ian Randle, 2006), 242.

35. Richard Dyer, *White: Essays on Race and Culture* (London: Routledge, 1997), 103.

36. Richard Dyer, "White," *Screen* 29 (Fall 1988): 44.

37. Nakamura, *Digitizing Race*.

38. Benjamin Wilson, Judy Hoffman, and Jamie Morgenstern, "Predictive Inequity in Object Detection," *arXiv preprint:1902.11097* (2019), 2–3.

39. John R. Feiner, John W. Severinghaus, and Philip E. Bickler, "Dark Skin Decreases the Accuracy of Pulse Oximeters at Low Oxygen Saturation: The Effects of Oximeter Probe Type and Gender," *Anesthesia & Analgesia* 105, no. 6 (2007): S18–S23.

40. Anna, C. Shcherbina, et al., "Accuracy in Wrist-Worn, Sensor-Based Measurements of Heart Rate and Energy Expenditure in a Diverse Cohort," *Journal of Personalized Medicine* 7, no. 2 (2017): 3–14.

41. For more discussion of the racial coding of technology and its violent outcomes, see Benjamin, *Race after Technology*.

42. Lorna Roth, "Looking at Shirley, the Ultimate Norm: Colour Balance, Image Technologies, and Cognitive Equity," *Canadian Journal of Communication* 34, no. 1 (2009): 111–136; Mary Ann Doane, "Screening the Avant-Garde Face," in *The Question of Gender*, eds. Judith Butler and Elizabeth Weed (Bloomington: Indiana University Press, 2011): 206–229; Genevieve Yue, "The China Girl on the Margins of Film," *October* 153 (2015): 96–116; Mulvin and Sterne, "Scenes from an Imaginary Country"; Susan Murray, *Bright Signals* (Durham, NC: Duke University Press, 2018).

43. Yue, "The China Girl."

44. Yue, "The China Girl," 99.

45. Coincidentally, Lena Forsén worked as a Shirley model as well as posing for *Playboy*. Linda Kinstler, "Finding Lena, the Patron Saint of JPEGs," *Wired* (January 31, 2019), https://www.wired.com/story/finding-lena-the-patron-saint-of-jpegs/.

46. Roth, "Looking at Shirley," 112.

47. Murray, *Bright Signals*; Jonathan Sterne and Dylan Mulvin, "The Low Acuity for Blue: Perceptual Technics and American Color Television," *Journal of Visual Culture* 13, no. 2 (2014): 118–138.

48. Mulvin and Sterne, "Scenes from an Imaginary Country."

49. The image, along with this quote, can be found in Gordon Comstock, "Jennifer in Paradise: The Story of the First Photoshopped Image," *The Guardian* (June 13, 2014), https://www .theguardian.com/artanddesign/photography-blog/2014/jun/13/photoshop-first-image -jennifer-in-paradise-photography-artefact-knoll-dullaart.

50. See also Philip W. Sewell, *Television in the Age of Radio: Modernity, Imagination, and the Making of a Medium* (New Brunswick, NJ: Rutgers University Press, 2014); Brian Winston, *Technologies of Seeing: Photography, Cinematography and Television* (London: British Film Institute, 1997); Murray, *Bright Signals*.

51. Dyer, *White*, 94.

52. Roth, *Looking at Shirley*; Dyer, *White*.

53. Roth, *Looking at Shirley*, 121–122 (emphasis in original).

54. Personal correspondence from David Myers.

55. Anna Lauren Hoffman, "Terms of Inclusion: Data, Discourse, Violence," *New Media & Society* (September 2020), https://doi.org/10.1177/1461444820958725.

56. bell hooks, "Eating the Other: Desire and Resistance," in *Media and Cultural Studies: Keyworks*, eds. Meenakshi Gigi Durham and Douglas Kellner (Malden, MA: Wiley, 2012), 308.

57. The images discussed here only scratch the surface of the ways that images are used as standards and stand-ins, which could be extended to include everything from ambiguous images used in psychological testing to the stock image industry; see Peter Galison, "Image of Self," in *Things That Talk* (Brooklyn: Zone Books), 257–294; and Paul Frosh, *The Image Factory* (Oxford, UK: Berg Publishers, 2003), respectively.

58. See *A Century of Excellence in Measurements, Standards, and Technology: A Chronicle of Selected NBS/NIST Publications 1901–2000*, ed. David R. Lide (Washington, DC: US Department of Congress, 2001).

59. Frank Rosenblatt, "The Perceptron: A Probabilistic Model for Information Storage and Organization in the Brain," *Psychological Review* 65, no. 6 (1958): 386–408.

60. Azriel Rosenfeld, "From Image Analysis to Computer Vision: An Annotated Bibliography, 1955–1979," *Computer Vision and Image Understanding* 84, no. 2 (2001): 298.

61. Janet Abbate, *Inventing the Internet* (Cambridge, MA: MIT Press, 1999), 36; Lisa Gitelman, *Always Already New* (Cambridge, MA: MIT Press, 2006); Fred Turner, *From Counterculture to Cyberculture: Stewart Brand, the Whole Earth Network, and the Rise of Digital Utopianism* (Chicago: University of Chicago Press, 2006).

62. Lawrence G. Roberts, "Picture Coding Using Pseudo-Random Noise," *IEEE Transactions on Information Theory* 8, no. 2 (1962): 145–154.

63. Because she was a child when she posed for these images, I am not naming her in this work or reproducing the images here.

64. William K. Pratt, "A Bibliography on Television Bandwidth Reduction Studies," *IEEE Transactions on Information Theory* 13, no. 1 (1967): 114–115.

65. Lawrence G. Roberts, *Machine Perception of Three-Dimensional Solids* (New York: Garland Publishing, 1980).

66. "TAKE ME, I'M YOURS: The Autobiography of SAIL," http://infolab.stanford.edu/pub /voy/museum/pictures/AIlab/SailFarewell.html. With thanks to Nathan Ensmenger for making me aware of this letter's existence.

67. Chris Garcia, "Robots Are a Few of My Favorite Things," *Computer History Museum*. June 17, 2015, https://computerhistory.org/blog/robots-are-a-few-of-my-favorite-things-by-chris -garcia/?key=robots-are-a-few-of-my-favorite-things-by-chris-garcia.

68. Because this reminiscence suggests that the model did not consent to being monitored by other men throughout the lab on a CCTV system, I am choosing not to provide a link to the images here.

69. Robin Lynch, "Man Scans: The Matter of Expertise in Art and Technology Histories," *RACAR* (Spring 2021, forthcoming).

70. "About SIPI," https://minghsiehece.usc.edu/groups-and-institutes/sipi/about/.

71. The IPTO was founded in 1962, at which point ARPA became a prime funder of computer science in the United States. SIPI, then called "USC-IPI," was funded by Contract number F08606–72–C-0008, Order number 1706 with ARPA's IPTO. As Amy Slaton and Janet Abbate note, the ARPANET node at USC, housed at the nearby Information Sciences Institute, was the "biggest and most heavily used ARPANET site." Amy Slaton and Janet Abbate, "The Hidden Lives of Standards: Technical Prescriptions and the Transformation of Work in America," in *Technologies of Power*, eds. Michael Thad Allen and Gabrielle Hecht (Cambridge, MA: MIT Press, 2001), 131.

72. William K. Pratt, *Digital Image Processing: PIKS Scientific Inside* (Hoboken, NJ: Wiley-Interscience, 2007), xvii.

73. "About SIPI."

74. William K. Pratt and Harry C. Andrews, Transform Processing and Coding of Images (Los Angeles, SIPI, 1969), 1.

75. Pratt and Andrews, Transform Processing.

76. William K. Pratt, *USCEE Report #411: Semi-annual Technical Report Covering Research Activity during the Period 3 August 1971 to 29 February 1972* (Los Angeles: SIPI, February 1972), i.

77. Ivan Sutherland, "Oral History Interview with Ivan Sutherland" (Minneapolis: Charles Babbage Institute, University of Minnesota Digital Conservancy, 1989), http://purl.umn .edu/107642 (emphasis added).

78. *A Century of Excellence in Measurements*.

79. The "Girl" image still appears in recent editions of Pratt's *Digital Image Processing* and *Introduction to Digital Image Processing* (Boca Raton, FL: Taylor & Francis, 2014).

80. William Pratt, *USCIPI Report #660: Semi-annual Technical Report Covering Research Activity During the Period 1 September 1975 to 31 March 1976* (Los Angeles: SIPI, March 1976).

81. This is as far as I can determine from reading every digitized SIPI report. It's possible that there are unavailable reports that would predate this publication.

82. Gitelman, *Always Already New*, 97; on the historiography of computing and the dominance of a "Silicon Valley Mythology," see Joy Lisi Rankin, *A People's History of Computing in the United States* (Cambridge, MA: Harvard University Press, 2018), 2.

83. Abbate, *Inventing the Internet*.

84. Abbate, *Inventing the Internet*, 175 (emphasis added).

85. Robert E. Kahn, "Oral History Interview with Robert E. Kahn," (Minneapolis: Charles Babbage Institute, University of Minnesota Digital Conservancy, 1989), http://purl.umn .edu/107380.

86. As Abbate notes in *Inventing the Internet*, IPTO leadership was very adept at construing a military justification for their work by often recharacterizing what was "research" and what was "development," depending on what they thought Congress wanted to hear.

87. Hanna Rose Shell, *Hide and Seek: Camouflage, Photography, and the Media of Reconnaissance* (New York: Zone Books, 2012).

88. Nowak, *Sex*, 173.

89. Nowak, *Sex*, 173.

90. Hutchinson, "Culture, Communication, and an Information Age Madonna," 1.

91. The full entry is located at "Lenna," *The Jargon File*, http://www.catb.org/jargon/html/L /lenna.html.

92. For instance, in 2012, researchers in Singapore received widespread attention for printing a copy of the Lena image that measured only 50 micrometers across, the smallest image ever printed. "Playboy Centrefold Photo Shrunk to Width of Human Hair," http://www.bbc .com/news/technology-19260550.

93. Linda Williams, *Hard Core: Power, Pleasure, and the "Frenzy of the Visible"* (Berkeley: University of California Press, 1989).

94. US Supreme Court, *Miller v. California*, 413 U.S. 15 (June 21, 1973).

95. Cynthia Enloe, *Maneuvers: The International Politics of Militarizing Women's Lives* (Berkeley: University of California Press, 2000).

CHAPTER 4

1. Sunny Bains, "Nude Image Creates Feelings of Exclusion," *Electronic Engineering Times* (May 26, 1997): 45.

2. On the concept of "professional vision," see chapter 3 of this book and Charles Goodwin, "Professional Vision," *American Anthropologist* 96, no. 3 (1994): 606–633.

3. USC-SIPI Image Database, http://sipi.usc.edu/database/.

4.   Jamie Hutchinson, "Culture, Communication, and an Information Age Madonna," *IEEE Professional Communication Society Newsletter* 45, no. 3 (2001): 1.

5.   Quoted in Howard Rheingold, *The Virtual Community: Homesteading on the Electronic Frontier* (Cambridge, MA: MIT Press, 2000), 136.

6.   See also Bruno Latour and Steve Woolgar, *Laboratory Life: The Construction of Scientific Facts* (Princeton, NJ: Princeton University Press, 1979).

7.   Hutchinson, "Culture, Communication, and an Information Age Madonna"; Peter Nowak, *Sex, Bombs, and Burgers: How War, Porn, and Fast Food Shaped Technology as We Know It* (New York: Viking).

8.   One of the earliest articles where the image is labeled with the name "Lena" concerned a technique for improving the quality of satellite television's NTSC-coded images. In this article, the image is described as "the woman in the hat (Lena)." Kerns H. Powers, "Techniques for Increasing the Picture Quality of NTSC Transmissions in Direct Satellite Broadcasting," *IEEE Journal on Selected Areas in Communications* 3, no. 1 (1985): 61.

9.   Brian J. Thompson, "Editorial: Copyright Problems," *Optical Engineering* 31, no. 1 (1992): 5.

10. Thompson, "Editorial," 5 (emphasis added).

11. The image appears on the covers of the July 1987 and April 1990 special issues on image processing, as well as appearing in articles published in those issues.

12. Thompson, "Editorial," 5 (emphasis in original).

13. Janelle Brown. "Playmate Meets Geeks Who Made Her a Net Star," *Wired* (May 1997), http://archive.wired.com/culture/lifestyle/news/1997/05/4000.

14. David C. Munson, "A Note on Lena," *IEEE Transactions On Image Processing* 5, no. 1 (1996): 3.

15. Munson, "A Note on Lena," 3.

16. In this tally, I counted total reproductions: if there was a sequence of four images showing four different steps of a process or contrasting different processes, that counted as four reproductions. I wasn't trying to count discrete tasks that used the Lena image, but rather to do a rough accounting of the image's dominance of the population of images in this one important journal.

17. On prototypical whiteness, see chapter 3 of this book and Simone Browne, *Dark Matters* (Durham, NC: Duke University Press, 2015); Richard Dyer, *White: Essays on Race and Culture* (London: Routledge, 1997); Lewis Gordon, "Is the Human a Teleological Suspension of Man?" in *After Man, Towards the Human: Critical Essays on Sylvia Wynter,* ed. Anthony Bogues (Kingston, Jamaica: Ian Randle, 2006), 237–257; Brian Winston, *Technologies of Seeing: Photography, Cinematography and Television* (London: British Film Institute, 1997).

18. Hutchinson, "Culture, Communication, and an Information Age Madonna," 5.

19. Clare Hemmings, *Why Stories Matter: The Political Grammar of Feminist Theory* (Durham, NC: Duke University Press, 2011).

20. Munson, "A Note on Lena," 3.

21. Munson, "A Note on Lena," 3.

22. Munson, "A Note on Lena," 3.

23. What the readership of *Image Processing* may not have known was that this controversy was playing out in the midst of a debate within feminist communities over pornography and sex work. The feminist "sex wars" or "porn wars" began in the 1980s and created schisms in American and European feminist politics. Antipornography feminists viewed porn and sexual practices like bondage and discipline, dominance and submission, sadism, and masochism (BDSM) as inescapably oppressive extensions of a patriarchal form of domination; meanwhile, pro-sex and sex-positive feminists fought for the recognition of sex work as legitimate and dignified labor and advocated for the liberatory and subversive potential of some forms of porn. However, this is something of an oversimplification. See Hemmings, *Why Stories Matter*; Rachel Corbman, "The Scholars and the Feminists: The Barnard Sex Conference and the History of the Institutionalization of Feminism," *Feminist Formations* 27, no. 3 (2015): 49–80.

24. Munson, "A Note on Lena," 3.

25. Munson, "A Note on Lena," 3 (emphasis added).

26. Rosenberg also maintains a website devoted to the Lena/Lenna image. http://www.lenna .org. As quoted in Hutchinson, "Culture," 6.

27. Wendy Hui Kyong Chun, *Control and Freedom: Power and Paranoia in the Age of Fiber Optics* (Cambridge, MA: MIT Press, 2006), 12–13.

28. Chun, *Control and Freedom*.

29. Cait McKinney, "Crisis Infrastructures: AIDS Activism Meets Internet Regulation," in *AIDS and the Distribution of Crises*, eds. Jih-Fei Cheng, Alexandra Juhasz, and Nishant Shahani (Durham, NC: Duke University Press, 2020), 162–182.

30. Amy Adele Hasinoff, *Sexting Panic: Rethinking Criminalization, Privacy, and Consent* (Urbana: University of Illinois Press, 2015), 130.

31. Sarah Banet-Weiser, *Empowered: Popular Feminism and Popular Misogyny* (Durham, NC: Duke University Press, 2018); Jessie Daniels, "Cloaked Websites: Propaganda, Cyber-racism and Epistemology in the Digital Era," *New Media & Society* 11, no. 5 (2009): 659–683.

32. *Reno v. American Civil Liberties Union*, 521 U.S. 844 (US Supreme Court, 1997).

33. Rebecca Tushnet, "Power without Responsibility: Intermediaries and the First Amendment," *George Washington Law Review* 76, no. 4 (2008): 986–1016.

34. Recording Industry Association of America (RIAA) (n.d.), "Resources & Learning," https://www.riaa.com/resources-learning/about-piracy/.

35. This was prior to the Digital Millennium Copyright Act (1998), which did provide some safe harbor rights to websites that hosted third-party content.

36. United States District Court, Northern District of Ohio (Eastern Division), *Playboy Enterprises Inc. v. Russ Hardenburgh, Inc.,* 982 F. Supp. 503 (November 25, 1997).

37. Although platforms today make claims about the scale and accuracy of their automated content moderation, a significant portion of the labor of determining these elements continues to be done manually (and note that image classification was a topic of the earliest studies in which the Lena image was used to test at SIPI). Sarah T. Roberts, *Behind the Screen* (New Haven, CT: Yale University Press, 2019).

38. Amy Hasinoff has shown how, in the context of the nonconsensual sharing of intimate images, property rights have become useful though problematic tools in the fight to protect privacy and reduce harm. Hasinoff, *Sexting Panic.*

39. Fred Turner, *From Counterculture to Cyberculture: Stewart Brand, the Whole Earth Network, and the Rise of Digital Utopianism* (Chicago: University of Chicago Press, 2006).

40. Judith Butler, *Bodies That Matter: On the Discursive Limits of "Sex"* (New York: Routledge, 1993), 139.

41. Martha Lampland and Susan Leigh Star, eds., *Standards and Their Stories* (Ithaca, NY: Cornell University Press, 2009); Geoffrey C. Bowker and Susan Leigh Star, *Sorting Things Out: Classification and Its Consequences* (Cambridge, MA: MIT Press, 1999).

42. Sara Ahmed, *Willful Subjects* (Durham, NC: Duke University Press, 2014), 143.

43. William K. Pratt, *Digital Image Processing: PIKS Scientific Inside* (Hoboken, NJ: John Wiley & Sons, 2007).

44. For instance, see Janet Abbate, *Recoding Gender: Women's Changing Participation in Computing* (Cambridge, MA: MIT Press, 2012); Amy Sue Bix, *Girls Coming to Tech!* (Cambridge, MA: MIT Press, 2014); Nathan Ensmenger, *The Computer Boys Take Over: Computers, Programmers, and the Politics of Technical Expertise* (Cambridge, MA: MIT Press, 2010); Jacob Gaboury, "A Queer History of Computing," *Rhizome,* February 19, 2013, https://rhizome.org/editorial/2013/feb/19/queer-computing-1/; Mar Hicks, *Programmed Inequality: How Britain Discarded Women Technologists and Lost Its Edge in Computing* (Cambridge, MA: MIT Press, 2017); Jennifer S. Light, "When Computers Were Women," *Technology and Culture* 40, no. 3 (1999): 455–483; Whitney Pow, "Outside of the Folder, the Box, the Archive" *ROMchip* 1, no. 1 (2019), https://romchip.org/index.php/romchip-journal/article/view/76; Christina Dunbar-Hester, *Hacking Diversity: The Politics of Inclusion in Open Technology Cultures* (Princeton, NJ: Princeton University Press, 2020).

45. Hicks, *Programmed Inequality*, 12.

46. Donna Riley and Gina L. Sciarra, "'You're All a Bunch of Fucking Feminists': Addressing the Perceived Conflict between Gender and Professional Identities Using the Montreal Massacre," *Proceedings of 36th ASEE/IEEE Frontiers in Education Conference* (2006): 19.

47. Wendy Hui Kyong Chun, "Unbearable Witness: Toward a Politics of Listening," *differences: A Journal of Feminist Cultural Studies,* 11, no. 1 (1999): 119.

48. Joan W. Scott, "Multiculturalism and the Politics of Identity," *October* 61 (1992): 12.

49. Ellen Spertus, "Why Are There So Few Female Computer Scientists?" MIT Artificial Intelligence Laboratory, 1991; Janet Cottrell, "I'm a Stranger Here Myself: A Consideration of Women in Computing," *ACM SIGUCCS User Services Conference* 20 (1992); Marianne Winslett, ed., *Final Report of the Committee on the Status of Women Graduate Students and Faculty in the College of Engineering,* University of Illinois at Urbana-Champaign, 1993; Karen A. Frenkel, "Women and Computing," *Communications of the ACM* 33, no. 11 (1990): 34–46; Barbara J. Grosz, ed., *Report on Women in the Sciences at Harvard,* Harvard University (1991); *Report of the MIT Committee on Family and Work,* Massachusetts Institute of Technology (1990).

50. Rebecca Slayton, "Revolution and Resistance: Rethinking Power in Computing History," *IEEE Annals of the History of Computing* 30, no. 1 (2008): 96–97; Dunbar-Hester, *Hacking Diversity.*

51. Spertus, "Why Are There So Few Female Computer Scientists?" 1.

52. Spertus, "Why Are There So Few Female Computer Scientists?"

53. It is important to note, however, that while there were (and continue to be) significant gender and racial disparities within the engineering and computer science programs at American and Canadian universities, the larger labor chain of technological production is heavily staffed by racialized women. Gender imbalance is a problem that afflicts elite universities and technology firms, but this focus can often conceal the larger systems of production and consumption, in which women and racialized populations manufacture computers and electronics, perform maintenance and repair work, and process e-waste. Dunbar-Hester, *Hacking Diversity*; Lisa Nakamura, "Indigenous Circuits: Navajo Women and the Racialization of Early Electronic Manufacture," *American Quarterly* 66, no. 4 (2014): 919–941.

54. Spertus, "Why Are There So Few Female Computer Scientists?" 24.

55. Spertus, "Why Are There So Few Female Computer Scientists?"; the Carnegie Mellon University report is quoted in Spertus, "Why Are There So Few Female Computer Scientists?" 25–26.

56. Banet-Weiser, *Empowered.*

57. Meredith Broussard *Artificial Unintelligence: How Computers Misunderstand the World* (Cambridge, MA: MIT Press, 2018), 69.

58. Broussard, *Artificial Unintelligence,* 69; As Fred Turner writes,

> The first *computer* hackers emerged at MIT in 1959. They were undergraduates who clustered around a giant TX-0 computer that had been built for defense research and then donated to MIT. Within several years, these undergraduates were joined by a variety of Cambridge-area teenagers

and MIT graduate students and began working with a series of computers donated by the Digital Equipment Corporation (DEC). By 1966 most of these young programmers gathered on the ninth floor of Technology Square, in Marvin Minsky's Artificial Intelligence (AI) Laboratory. (*From Counterculture to Cyberculture*, 133 [emphasis in original]).

See also Gabriella Coleman, *Coding Freedom* (Princeton, NJ: Princeton University Press, 2012); Dunbar-Hester, *Hacking Diversity*.

59. For more on the history of the Media Lab and its culture, see Molly Wright Steenson, *Architectural Intelligence: How Designers and Architects Created the Digital Landscape* (Cambridge, MA: MIT Press, 2017).

60. Minsky died in 2016. In a recently unsealed deposition, Virginia Roberts Giuffre accused Jeffrey Epstein and Ghislaine Maxwell of trafficking her as a minor to Epstein's private island to have coerced sex. She names Marvin Minsky as one of the people she was forced to have sex with. When the accusations came to light in 2019, a Media Lab member and founder of the Free Software Foundation, Richard Stallman, downplayed the accusations against Minsky. In an email to the CSAIL mailing list, Stallman wrote that he took exception to the use of the word "assault" to describe the accusation against Minsky: "The word 'assaulting' presumes that he applied force or violence, in some unspecified way, but the article itself says no such thing. Only that they had sex." Stallman's email was exposed in a *Medium* post by an MIT graduate named Selam Jie Gano. Stallman was forced to step down for his comments, which were taken as not only insufficiently compassionate and factually wrong, but emblematic of a toxic and abusive culture within the CSAIL and Media Lab communities. Gano's *Medium* post publishing the email begins, aptly, "I'm writing this because I'm too angry to work." Selam Jie Gano, "Remove Richard Stallman," *Medium,* https://medium.com/@selamjie /remove-richard-stallman-fec6ec210794; Victoria Bekiempis, "MIT Scientist Resigns over Emails Discussing Academic Linked to Epstein," *The Guardian*, September 17, 2019, https:// www.theguardian.com/education/2019/sep/17/mit-scientist-emails-epstein.

61. Frenkel notes that this passage, one of many about MIT in the issue, is from Jennifer Tidwell's unpublished paper on MIT's "Terminal Garden" written in the spring of 1990. Frenkel, "Women and Computing," 36–37.

62. Carrie Rentschler, "Witnessing: US Citizenship and the Vicarious Experience of Suffering" *Media, Culture & Society* 26, no. 2 (2004): 296–304.

63. Amia Srinivasan, "The Aptness of Anger," *Journal of Political Philosophy* 26, no. 2 (2018): 127.

64. Srinivasan, "Aptness," 136.

65. Kenneth Burke, *Permanence and Change: An Anatomy of Purpose* (Los Altos, CA: Hermes Publications, 1954), 18.

66. Thorstein Veblen, *The Instinct of Workmanship and the State of the Industrial Arts* (New York: Macmillan, 1914).

67. Burke, *Permanence and Change*, 18.

68. Luce Irigaray, *Speculum of the Other Woman* (Ithaca, NY: Cornell University Press, 1985), 13 (emphasis in original).

69. Michael T. Eismann, "Farewell, Lena," *Optical Engineering* 57, no. 12 (2018): 120101.

70. "A Note on the Lena Image," *Nature Nanotechnology* 13, no. 12 (2018): 1087.

71. A sonnet written to the Lena image, by Thomas Colthurst, was published online in the early 2000s and since then has been reposted widely. The original is available at http://thomaswc.com/poems.html.

72. A 2016 exhibit at the Vancouver Art Gallery called "Mashup: The Birth of Modern Culture" features the Lena image in an artwork by Amber Frid-Jimenez. In this piece, *This Is Not a Test* (2016), the full centerfold is wrapped around a 3D triangular sculpture and is overlaid with quotes from feminist authors. Another artwork, by Trevor Paglen, uses the full centerfold—though he has removed the *Playboy* watermark—for a piece titled *Lenna: Empress of Invisible Images, Queen of the Internet* (2017).

73. This video is viewable on the artist's website: http://www.jamieallen.com/killinglena/.

74. Sara Ahmed, *Queer Phenomenology: Orientations, Objects, Others* (Durham, NC: Duke University Press, 2006).

75. John Berger, *Ways of Seeing* (London: Penguin Books, 1972).

CHAPTER 5

1. Kevin Ferguson, "How Do You Teach Medical Students Bedside Manner? Hire an Actor," *Off-Ramp*, August 20, 2015, radio program, https://www.scpr.org/programs/offramp/2015/08/20/44184/how-do-you-teach-medical-students-bedside-manner-h/ (emphasis added).

2. Janelle S. Taylor, "The Moral Aesthetics of Simulated Suffering in Standardized Patient Performances," *Culture, Medicine, and Psychiatry* 35, no. 2 (2011): 138 (emphasis added).

3. Here, I am referring to the work of Arlie Russell Hochschild, who coined the term "emotional labor" during her study of airline attendants and bill collectors, though the term is now widely used in a range of disciplines and denotes a much wider set of practices than Hochschild's originally restrained scope. She intentionally excluded doctors from her original study of those susceptible to the regimes of emotional labor because their emotions were not supervised. However, since the time she first published her book, *The Managed Heart*, this kind of supervision has become a necessary component of medical education and licensing, as seen in the standardized patient program. Arlie Russell Hochschild, *The Managed Heart: Commercialization of Human Feeling*, 2nd ed. (Berkeley: University of California Press, 2012).

4. This has been the case in Canada since 1993 and the United States since 2004. The first physicians who were required to pass the SP test were foreign-trained physicians trying to practice in either country. Before it seemed necessary or acceptable to require domestically trained physicians to pass a test using SPs or to impose it as a licensing requirement, it was field-tested on "outsiders" of the Canadian and American medical systems. Brian David

Hodges and Nancy McNaughton, "Who Should Be an OSCE Examiner?" *Academic Psychiatry* 33, no. 4 (2009): 282–284; L. Stephen Jacyna and Stephen T. Casper, *The Neurological Patient in History* (Rochester, NY: University of Rochester Press, 2012); Roy Porter, "The Patient's View: Doing Medical History from Below," *Theory and Society* 14, no. 2 (1985): 175–198.

5.   Sara Ahmed, *The Promise of Happiness* (Durham, NC: Duke University Press, 2010); Patricia T. Clough, "The Affective Turn: Political Economy, Biomedia and Bodies," *Theory, Culture & Society, 25*, no. 1 (2008): 1–22; Ruth Leys, *The Ascent of Affect* (Chicago: University of Chicago Press, 2017).

6.   Hodges and McNaughton, "Who Should Be an OSCE Examiner?" 282.

7.   Adam I. Levine and Mark H. Swartz, "Standardized Patients: The 'Other' Simulation," *Journal of Critical Care* 23, no. 2 (2008): 179–184.

8.   Physicians proceed through twelve examinations, with an interview and physical examination of a patient that lasts fifteen minutes and a ten-minute station where the examinee writes a record of the SP's history and physical findings.

9.   Donald E. Melnick, Gerard F. Dillon, and David B. Swanson, "Medical Licensing Examinations in the United States," *Journal of Dental Education* 66, no. 5 (2002): 595–599.

10.   Howard S. Barrows, Paul R. Patek, and Stephen Abrahamson, "Introduction of the Living Human Body in Freshman Gross Anatomy," *British Journal of Medical Education* 2, no. 1 (1968): 33–35.

11.   Howard Barrows and Stephen Abrahamson, "The Programmed Patient: A Technique for Appraising Student Performance in Clinical Neurology," *Journal of Medical Education* 39 (1964): 802–805.

12.   SPs and related persons have been given many other names as well: "patient instructor," "patient educator," "professional patient," "surrogate patient," and "teaching associate." Howard S. Barrows, "Simulated Patients in Medical Training," *Canadian Medical Association Journal* 98 (1968): 674–676; Howard S. Barrows, *Simulated Patients (Programmed Patients): The Development and Use of a New Technique in Medical Education* (Springfield, IL: Charles C. Thomas, 1971).

13.   Peggy Wallace, "Following the Threads of an Innovation: The History of Standardized Patients in Medical Education," *Caduceus* 13, no. 2 (1997): 6.

14.   C. Donald Combs, "Humans as Models," in *Modeling and Simulation in the Medical and Health Sciences*, eds. John A. Sokolowski and Catherine M. Banks (Hoboken, NJ: John Wiley & Sons, Inc., 2011), 92.

15.   Michel Foucault, *The Birth of the Clinic*, trans. A. M. Sheridan (London: Routledge, 2003[1973]), 122.

16.   Hsuan L. Hsu and Martha Lincoln, "Biopower, Bodies . . . the Exhibition, and the Spectacle of Public Health," *Discourse*, 29, no. 1 (2007): 23.

17. John Forrester, "If *p* Then What? Thinking in Cases," *History of the Human Sciences* 9, no. 3 (1996): 1–25.

18. University of Texas Medical Branch, "Template for Standardized Patient Script," http://www.utmb.edu/ocs/FacDevTools/script-template.asp.

19. Baylor College of Medicine, "Standardized Patient Script Example: Back Pain Script," https://www.bcm.edu/education/schools/medical-school/programs/standardized-patient-program/become-a-standardized-patient/script-example.

20. Rachel Hall, *The Transparent Traveler* (Durham, NC: Duke University Press, 2015), 4.

21. The MIRS document is adapted from the Arizona Interview Rating Scale (ACIR), developed by Paula Stillman, and it is written to reflect the 2001 Kalamazoo Consensus on "Essential Elements of Communication in Medical Encounters." Gregory Makoul, "Essential Elements of Communication in Medical Encounters: The Kalamazoo Consensus Statement," *Academic Medicine* 76, no. 4 (2001): 390–393.

22. Elaine Scarry, *The Body in Pain: The Making and Unmaking of the World* (New York: Oxford University Press, 1985), 12.

23. Scarry writes that "if property (as well as the ways in which property can be jeopardized) were easier to describe than bodily disability (as well as the ways in which a disabled person can be jeopardized), then one would not be astonished to discover that a society had developed sophisticated procedures for protecting 'property rights' long before it had succeeded in formulating the concept of 'the rights of the handicapped.'" *The Body in Pain*, 12.

24. Scarry, *The Body in Pain*, 13.

25. Ronald Melzack, "The McGill Pain Questionnaire: Major Properties and Scoring Methods," *Pain* 1, no. 3 (1975): 277.

26. Lochlann Jain, *Injury: The Politics of Product Design and Safety Law in the United States* (Princeton, NJ: Princeton University Press, 2006); Michelle Murphy, *Sick Building Syndrome and the Problem of Uncertainty: Environmental Politics, Technoscience, and Women Workers* (Durham, NC: Duke University Press, 2006).

27. Kelly M. Hoffman, Sophie Trawalter, Jordan R. Axt, and M. Norman Oliver, "Racial Bias in Pain Assessment and Treatment Recommendations, and False Beliefs about Biological Differences between Blacks and Whites," *Proceedings of the National Academy of Sciences* 113, no. 16 (2016): 4298.

28. Hoffman et al., "Racial Bias," 4297–4298.

29. Hoffman et al., "Racial Bias," 4300.

30. Soraya Chemaly, *Rage Becomes Her* (New York: Atria, 2018); Francis B. Nyamnjoh, "Black Pain Matters: Down with Rhodes," *Pax Academica* 1, no. 2 (2015): 47–70.

31. Elena Ruíz, "Cultural Gaslighting," *Hypatia* 35, no. 4 (2020): 687–713; Harriet A. Washington, *Medical Apartheid: The Dark History of Medical Experimentation on Black Americans from Colonial Times to the Present* (New York: Doubleday Books, 2006).

32. Ruíz, "Cultural Gaslighting."

33. Angelique M. Davis and Rose Ernst, "Racial Gaslighting," *Politics, Groups, and Identities* 7, no. 4 (2019): 761; Ruíz, "Cultural Gaslighting."

34. Barrows and Abrahamson, "The Programmed Patient," 803 (emphasis added).

35. Barrows and Abrahamson, "The Programmed Patient," 803 (emphasis added).

36. Erving Goffman, *Stigma: Notes on the Management of Spoiled Identity* (Englewood Cliffs, NJ: Prentice Hall, 1962), 5.

37. Rosemary Garland Thomson, *Extraordinary Bodies: Figuring Physical Disability in American Culture and Literature* (New York: Columbia University Press, 1997), 8; see also Aimi Hamraie, *Building Access: Universal Design and the Politics of Disability* (Minneapolis: University of Minnesota Press, 2017).

38. Robert McRuer, "Compulsory Able-Bodiedness and Queer/Disabled Existence," in *The Disability Studies Reader*, 3rd ed., ed. Lennard Davis (London: Routledge, 2010): 383–392.

39. Lennard Davis, *Enforcing Normalcy: Disability, Deafness, and the Body* (London: Verso, 1995), 2.

40. Meryl Alper, *Giving Voice* (Cambridge, MA: MIT Press, 2017); Hamraie, *Building Access*; Alison Kafer, *Feminist, Queer, Crip* (Bloomington; Indiana University Press, 2013).

41. Colin Barnes, "Understanding the Social Model of Disability," in *Routledge Handbook of Disability Studies*, eds. Nick Watson, Alan Roulstone, and Carol Thomas (London: Routledge, 2012), 12–29.

42. Barnes, "Understanding," 18.

43. Kafer, *Feminist, Queer, Crip*; Tom Shakespeare and Nicholas Watson, "The Social Model of Disability: An Outdated Ideology?" *Research in Social Science and Disability* 2 (2002): 9–28; Jonathan Sterne, "Ballad of the Dork-o-phone: Towards a Crip Vocal Technoscience," *Journal of Interdisciplinary Voice Studies* 4, no. 2 (2019): 179–189; Shelley Tremain, "On the Subject of Impairment," in *Disability/Postmodernity: Embodying Disability Theory*, eds. Mairian Corker and Tom Shakespeare (New York: Bloomsbury, 2002); Susan Wendell, *The Rejected Body: Feminist Philosophical Reflections on Disability* (New York: Routledge, 1996).

44. Kafer, *Feminist, Queer, Crip*, 9.

45. Kafer, *Feminist, Queer, Crip*.

46. Granola bars and mints are also the things I carried with me while interviewing for jobs. It is possible that they are a standard part of the equipment that many people use to maintain their bodies and minimize the discomfort that those bodies might compel in other people. Leslie Jamison, *The Empathy Exams* (Minneapolis: Graywolf Press, 2014), 3.

47. Louise Aronson, "Examining Empathy," *The Lancet* 384, no. 9937 (2014): 16.

48. Aronson, "Examining Empathy," 16.

49. Constantin Stanislavski, *An Actor Prepares* (London: Routledge, 1989), 12.

50. David Mendel, *Proper Doctoring: A Book for Patients and Their Doctors* (New York: New York Review Books Classics, 2013[1984]), 5.

51. Mendel, *Proper Doctoring*, 9.

52. Mendel, *Proper Doctoring*, 19.

53. Mendel, *Proper Doctoring*, 20.

54. Rebecca Schneider, *Performing Remains: Art and War in Times of Theatrical Reenactment* (London: Routledge, 2011), 41.

55. Schneider, *Performing Remains*, 41 (emphasis in original).

56. Dylan Mulvin, "Media Prophylaxis: Night Modes and the Politics of Preventing Harm," *Information & Culture* 53, no. 2 (2018): 175–202.

57. Richard Schechner, *Between Theatre and Anthropology* (Philadelphia: University of Pennsylvania Press 1985), 97.

58. Taylor, "The Moral Aesthetics."

59. Christopher Pearce and Steve Trumble, "Computers Can't Listen: Algorithmic Logic Meets Patient Centredness," *Australian Family Physician* 35, no. 6 (2006).

60. Barrows, "Simulated Patients in Medical Training,"; Howard S. Barrows, Paul R. Patek, and Stephen Abrahamson, "Introduction of the Living Human Body in Freshman Gross Anatomy," *Medical Education* 2, no. 1 (1968): 33–35.

61. Howard S. Barrows, "An Overview of the Uses of Standardized Patients for Teaching and Evaluating Clinical Skills," *Academic Medicine* 68 no. 6 (1993): 444.

62. Taylor, "The Moral Aesthetics," 155.

63. Tobin Siebers, "Disability as Masquerade," *Literature and Medicine* 23, no. 1 (2004): 18.

64. Sasha Constanza-Chock, *Design Justice* (Cambridge, MA: MIT Press, 2020); Kafer, *Feminist, Queer, Crip*.

65. Anne Herrmann-Werner, Teresa Loda, Lisa M. Wiesner, Rebecca Sarah Erschens, Florian Junne, and Stephan Zipfel, "Is an Obesity Simulation Suit in an Undergraduate Medical Communication Class a Valuable Teaching Tool? A Cross-sectional Proof of Concept Study," *BJM Open* 9 (2019): e029738.

66. Herrmann-Werner et al. "Obesity Simulation Suit."

67. Linda Long-Bellil et al., "Teaching Medical Students about Disability: The Use of Standardized Patients," *Academic Medicine* 86, no 9 (2011): 1166.

68. Long-Bellil et al., "Teaching Medical Students about Disability," 1166.

69. Barrows, "Overview," 446.

70. Wallace, "Following the Threads."

71. "Models Who Imitate Patients: Paradise for Medical Students," *San Francisco Chronicle*, September 28, 1965.

72. Barrows, *Simulated Patients*, 15.

73. Barrows, *Simulated Patients*, 14.

74. Barrows, *Simulated Patients*, 15.

75. Barrows, "Overview," 444 (emphasis added).

76. This was precisely the context in which Kramer and his friend Mickey became SPs in an episode of the television program *Seinfeld*. Kramer was a quintessential gig worker and (sometimes) actor. One of the jokes in the Seinfeld episode, "The Burning," in which Kramer must perform as a patient with gonorrhea, involves Kramer and Mickey escalating their SP performances to the point of absurdity and competing for the "best" diseases to perform. Barrows, *Simulated Patients*; Emily Cegielski, "For Actors, Pretending to Be Sick Can Pay Off," Backstage.com, http://www.backstage.com/news/for-actors-pretending-to-be-sick-can-pay-off/.

77. "Models Who Imitate Patients."

78. Lisa D. Howley, Gayle Gliva-McConvey, Judy Thornton, and Association of Standardized Patient Educators, "Standardized Patient Practices: Initial Report on the Survey of US and Canadian Medical Schools," *Medical Education Online* 14 (2009): doi: 10.3885/meo.2009. F0000208.

79. Lena H. Sun, "Demand Is High for Pretend Patients," *Washington Post*, October 14, 2011, A1.

80. See https://health.usf.edu.

81. Temple University School of Medicine, "Standardized Patient Program: Questions and Answers about Working as a Standardized Patient for Temple University School of Medicine" (2013): https://medicine.temple.edu/sites/medicine/files/files/FAQs.pdf.

82. Richard Terry, Erik Hiester, and Gary D. James, "The Use of Standardized Patients to Evaluate Family Medicine Resident Decision Making," *Family Medicine* 39, no. 4 (2007): 263.

83. Taylor, "The Moral Aesthetics."

84. Taylor, "The Moral Aesthetics," 156.

85. Greg Downey, "Making Media Work: Time, Space, Identity, and Labor in the Analysis of Information and Communication Infrastructures," in *Media Technologies: Essays on Communication, Materiality, and Society*, eds. Tarleton Gillespie, Pablo Boczkowski and Kirsten Foot (Cambridge, MA: MIT Press, 2014), 164.

86. Howard S. Barrows, *Simulated Patient Training: Acute Paralysis of Both Legs in a Young Woman* (Chapel Hill, NC: Health Sciences Consortium, 1988).

87. Barrows's original simulation, for a patient whom he gave the name "Patty Dugger," was also paraplegic, and in an early newspaper article about the program, it mentions that

Barrows's first patients were trained to simulate (presumably among other things) paralysis, loss of sensation, blindness, and abnormal reflexes. Wallace, "Following the Threads"; "Models Who Imitate Patients."

CHAPTER 6

1. Steven Jackson, "Rethinking Repair," in *Media Technologies: Essays on Communication, Materiality, and Society* (Cambridge, MA: MIT Press, 2014), 221–239.

2. Sarah Ganz Blythe and Edward D. Powers, *Looking at Dada* (New York: Museum of Modern Art, 2006), 52.

3. Marcel Duchamp, *The Writings of Marcel Duchamp* (New York: Da Capo Press, 1989), 22.

4. Herbert Molderings, *Duchamp and the Aesthetics of Chance: Art as Experiment* (New York: Columbia University Press, 2010), 2.

5. Ken Alder, *The Measure of All Things: The Seven-Year Odyssey and Hidden Error That Transformed the World* (New York: Free Press, 2002).

6. Matthew Fuller approaches "media ecologies" through standard objects—such as the shipping container and packet switching protocols—and he understands a standard to be a continuum with itself at one pole and total disorder at the other. *Media Ecologies: Materialist Energies in Art and Technoculture* (Cambridge, MA: MIT Press, 2005).

7. David Turnbull, "The Ad Hoc Collective Work of Building Gothic Cathedrals with Templates, String, and Geometry," *Science, Technology, & Human Values* 18, no. 3 (1993): 315–340.

8. Lawrence Busch, *Standards: Recipes for Reality* (Cambridge, MA: MIT Press, 2011).

9. Michael Baxandall, *Painting and Experience in Fifteenth-Century Italy: A Primer in the Social History of Pictorial Style* (Oxford: Oxford University Press, 1988), 86–89.

10. Jacob Gaboury, "Image Objects: An Archaeology of 3D Computer Graphics, 1965–1979," PhD dissertation (New York University, 2015); Ann-Sophie Lehmann, "Taking the Lid off the Utah Teapot towards a Material Analysis of Computer Graphics," *Zeitschrift für Medien-und Kulturforschung* 2012, no. 1 (2012): 169–184.

11. As Lehmann states about the Utah Teapot in "Taking the Lid off the Utah Teapot," 176,

    In the white, male, and mostly bearded scientific community that produced the first computer graphics in the 1970s, it was an object not from the lab but from the home, a different, yet familiar thing, blending the efficiency of mathematics with the cosiness of a warm cup of tea. The teapot thus crossed boundaries between genders and cultures, but also between art and science, as it served to develop and express visual creativity in a scientific environment, or joined the left and right side of the brain, as James Blinn once put it.

12. Hito Steyerl, *How Not to Be Seen: A Fucking Didactic Educational .MOV File*, 2013, video, Museum of Modern Art, New York, https://www.moma.org/collection/works/181784.

13. Christopher Williams, *Kodak Three Point Reflection Guide © 1968 Eastman Kodak Company, 1968 (Meiko laughing), Vancouver, B.C., April 6, 2005*, Whitney Museum of American Art, New York, https://whitney.org/collection/works/27531.

14. Ken Gewerts, "A Bevy of Unknown Beauties," *Harvard Gazette*, July 21, 2005, http://news.harvard.edu/gazette/2005/07.21/00-girls.html.

15. I attended this lecture myself, and both Raven and her audience focused on the formal simplicity of test materials as their most compelling aspect.

16. Adam Broomberg and Oliver Chanarin, *To Photograph the Details of a Dark Horse in Low Light*, Gallery TPW R&D and Scotiabank Contact Photography Festival, 2013, http://gallerytpw.ca/rd/broomberg-chanarin/billboard-locations/.

17. Robin Lynch, "Man Scans: The Matter of Expertise in Art and Technology Histories," *RACAR* (Spring 2021, forthcoming).

18. Genevieve Yue has compiled a long list of appropriations of test images in contemporary art. See Genevieve Yue, "The China Girl on the Margins of Film," *October*, no. 153 (2015), 96–116; Jonathan Sterne, *MP3: The Meaning of a Format* (Durham, NC: Duke University Press, 2012).

19. See Ryan Maguire, "The Ghost in the MP3," 2014, http://ryanmaguiremusic.com/theghostinthemp3.html.

20. Howard S. Barrows, "Simulation in Medical Education," *Caduceus* 13, no. 2 (1997): 4.

21. On other recent artistic appropriations of computer vision, see Jill Walker Rettberg, "Machine Vision as Viewed through Art: Hostile Other or Part of Ourselves?" Paper presented at Post-Screen Festival: PSF2016, Lisbon, November 17–18, 2016.

22. Paul Du Gay, Stuart Hall, Linda Janes, Anders Koed Madsen, Hugh Mackay, and Keith Negus, *Doing Cultural Studies: The Story of the Sony Walkman* (London: SAGE, 2013).

23. Friedrich Kittler, *Gramophone, Film, Typewriter* (Palo Alto, CA: Stanford University Press, 1999); Friedrich Kittler, *Optical Media* (Cambridge, UK: Polity, 2010).

24. Geoffrey C. Bowker, *Science on the Run Information Management and Industrial Geophysics at Schlumberger, 1920–1940* (Cambridge, MA: MIT Press, 1994), 10–14.

25. Susan Leigh Star and Karen Ruhleder, "Steps toward an Ecology of Infrastructure: Design and Access for Large Information Spaces," *Information Systems Research Information Systems Research* 7, no. 1 (1996): 111–134.

26. Geoffrey C. Bowker, Karen Baker, Florence Millerand, and David Ribes, "Toward Information Infrastructure Studies: Ways of Knowing in a Networked Environment," in *International Handbook of Internet Research*, ed. Jeremy Hunsinger, Lisbeth Klastrup, and Matthew M. Allen (Dordrecht, Netherlands: Springer, 2010); Geoffrey C. Bowker and Susan Leigh Star, *Sorting Things Out* (Cambridge, MA: MIT Press, 1999); Paul N. Edwards, *A Vast Machine: Computer Models, Climate Data, and the Politics of Global Warming* (Cambridge, MA: MIT Press, 2010); Sterne, *MP3*.

27. Star and Ruhleder, "Steps toward an Ecology of Infrastructure," 113 (emphasis added). Note, as well, that these items are drawn from a longer list of the characteristics of infrastructure.

28. Bowker et al., "Toward Information Infrastructure Studies."

29. Lisa Gitelman, *Always Already New* (Cambridge, MA: MIT Press, 2006), 7.

30. Nick Montfort and Ian Bogost, *Racing the Beam: the Atari Video Computer System* (Cambridge, MA: MIT Press, 2009), 2–3.

31. Tarleton Gillespie, *Wired Shut: Copyright and the Shape of Digital Culture* (Cambridge, MA: MIT Press, 2007), 93–94.

32. Mikkel Flyverbom, *The Digital Prism* (Cambridge: Cambridge University Press, 2019); Mike Ananny and Kate Crawford, "Seeing without Knowing: Limitations of the Transparency Ideal and Its Application to Algorithmic Accountability," *New Media & Society* 20, no. 3 (2018): 973–989.

33. Flyverbom, *Digital Prism*, 97.

34. Martha Lampland and Susan Leigh Star, *Standards and Their Stories* (Ithaca, NY: Cornell University Press, 2009), 17 (emphasis added).

35. Mentions of an infrastructure crisis in the United States started to gather in the 1980s. In subsequent decades they expanded and multiplied exponentially. In an early comment on this emergent pattern, Heywood T. Sanders expressed rare skepticism on the topic: "What Infrastructure Crisis?" *Public Interest*, no. 110 (1993): 3–18.

36. Star and Ruhleder, "Steps toward an Ecology of Infrastructure"; William J. Rankin, "Infrastructure and the International Governance of Economic Development, 1950–1965," in *Internationalization of Infrastructures*, eds. Jean-François Auger, Jan Jaap Bouma, and Rolf Künneke (Delft, Netherlands: Delft University of Technology, 2009): 61–75.

37. Jennifer Holt and Patrick Vonderau, "Where the Internet Lives," in *Signal Traffic* (Urbana: University of Illinois Press, 2015), 71–93; Mél Hogan, "Facebook Data Storage Centers as the Archive's Underbelly," *Television & New Media* 16, no. 1 (2015): 3–18; Louise Amoore, *Cloud Ethics: Algorithms and the Attributes of Ourselves and Others* (Durham, NC: Duke University Press, 2020).

38. Kregg Hetherington, "Surveying the Future Perfect: Anthropology, Development and the Promise of Infrastructure," in *Infrastructures and Social Complexity: A Companion*, eds. Penelope Harvey, Casper Bruun Jensen, and Atsuro Morita (London: Routledge, 2016): 42.

39. Shaylih Muehlmann, "Clandestine Infrastructures: Illicit Connectivities in the US-Mexico Borderlands," in *Infrastructure, Environment, and Life in the Anthropocene*, ed. Kregg Hetherington (Durham, NC: Duke University Press, 2019): 45–65; Brian Larkin, "The Politics and Poetics of Infrastructure," *Annual Review of Anthropology* 42, no. 1 (2013): 327–343; Lauren Berlant, "The Commons: Infrastructures for Troubling Times," *Environment and Planning D: Society and Space* 34, no. 3 (2016): 393–419; Megan Finn, *Documenting*

Aftermath: Information Infrastructures in the Wake of Disasters (Cambridge, MA: MIT Press, 2018); Marisa Elena Duarte, *Network Sovereignty: Building the Internet across Indian Country* (Seattle: University of Washington Press, 2017).

40. Gabrielle Hecht, "Interscalar Vehicles for an African Anthropocene: On Waste, Temporality, and Violence," *Cultural Anthropology* 33, no. 1 (2018): 109–141.

41. Paul N. Edwards, "Infrastructure and Modernity: Force, Time, and Social Organization in the History of Sociotechnical Systems," in *The History of Sociotechnical Systems: Modernity and Technology*, eds. Thomas Misa, Philip Brey and Andrew Feenberg (Cambridge, MA: MIT Press): 185–226.

42. Thomas Pachirat examines the "politics of sight" involved in the managed invisibility of the US meat industry and its slaughterhouses. For a different approach, see the issue of *Postcolonial Studies* on toilets and the politics of modernity and transparency. Thomas Pachirat, *Every Twelve Seconds* (New Haven, CT: Yale University Press, 2011); Michael Dutton, Sanjay Seth, and Leela Gandhi, "Plumbing the Depths: Toilets, Transparency, and Modernity," *Postcolonial Studies* 5, no. 2 (2002): 137–142.

43. Jas Rault, "Tricks of Transparency in Colonial Modernity," Digital Research Ethics Collaboratory (DREC), http://www.drecollab.org/tricks-of-transparency/.

44. Armond R. Towns, "Toward a Black Media Philosophy," *Cultural Studies* 34 (2020): 853.

45. Rachel Hall, *The Transparent Traveler* (Durham, NC: Duke University Press, 2015).

46. Shannon Mattern, "Maintenance and Care," *Places Journal,* November 2018, https://placesjournal.org/article/maintenance-and-care/.

47. Ingrid Burrington, *Networks of New York: An Illustrated Field Guide to Urban Internet Infrastructure* (Brooklyn: Melville House Printing, 2016).

48. Burrington, *Networks of New York.*

49. Craig Robertson, *The Passport in America: The History of a Document* (Oxford and New York: Oxford University Press, 2010).

50. Carolyn Steedman, "Something She Called a Fever: Michelet, Derrida, and Dust," *American Historical Review* 106, no. 4 (2001): 1163.

51. Sara Ahmed, *Queer Phenomenology: Orientations, Objects, Others* (Durham, NC: Duke University Press, 2006).

52. Judith Butler, *Bodies That Matter: On the Discursive Limits of "Sex"* (New York: Routledge, 1993), ix.

Bibliography

Abbate, Janet. *Inventing the Internet*. Cambridge, MA: MIT Press, 1999.

Abbate, Janet. *Recoding Gender: Women's Changing Participation in Computing*. Cambridge, MA: MIT Press, 2012.

Abraham, Katharine G., John S. Greenlees, and Brent R. Moulton. "Working to Improve the Consumer Price Index." *Journal of Economic Perspectives* 12, no. 1 (1998): 27–36.

Adams, J. Christian. *A Primer on "Motor Voter": Corrupted Voter Rolls and the Justice Department's Selective Failure to Enforce Federal Mandates*. Heritage Foundation (September 25, 2014). https://www.heritage.org/election-integrity/report/primer-motor-voter-corrupted-voter-rolls-and-the-justice-departments.

Agostinho, Daniela, and Nanna Bonde Thylstrup. "'If Truth Was a Woman': Leaky Infrastructures and the Gender Politics of Truth-Telling." *Ephemera* 19, no. 4 (2019): 745–775.

Ahmed, Sara. *The Promise of Happiness*. Durham, NC: Duke University Press, 2010.

Ahmed, Sara. *Queer Phenomenology: Orientations, Objects, Others*. Durham, NC: Duke University Press, 2006.

Ahmed, Sara. *Strange Encounters: Embodied Others in Post-Coloniality*. London: Routledge, 2000.

Ahmed, Sara. *Willful Subjects*. Durham, NC: Duke University Press, 2014.

Ajunwa, Ifeoma, Kate Crawford, and Jason Schultz. "Limitless Worker Surveillance." *California Law Review* 105, no. 3 (2017): 735–776.

Akrich, Madeline. "The De-Scription of Technical Objects." In *Shaping Technology/Building Society: Studies in Sociotechnical Change*, edited by Wiebe E. Bijker and John Law, 205–224. Cambridge, MA: MIT Press, 1992.

Albert, Kendra, Emily Armbruster, Elizabeth Brundige, Elizabeth Denning, Kimberly Kim, Lorelei Lee, et al. "FOSTA in Legal Context" (July 30, 2020). https://papers.ssrn.com/sol3/papers.cfm?abstract_id=3663898.

Alder, Ken. "Making Things the Same." *Social Studies of Science* 28, no. 4 (1998): 499–545.

Alder, Ken. *The Measure of All Things: The Seven-Year Odyssey and Hidden Error That Transformed the World*. New York: Free Press, 2002.

Alder, Ken. "Revolution to Measure: The Political Economy of the Metric System in France." In *The Values of Precision*, edited by M. Norton Wise, 39–71. Princeton, NJ: Princeton University Press, 1995.

Alper, Meryl. *Giving Voice: Mobile Communication, Disability, and Inequality*. Cambridge, MA: MIT Press, 2017.

Amoore, Louise. *Cloud Ethics: Algorithms and the Attributes of Ourselves and Others*. Durham, NC: Duke University Press, 2020.

Ananny, Mike, and Kate Crawford. "Seeing without Knowing: Limitations of the Transparency Ideal and Its Application to Algorithmic Accountability." *New Media & Society* 20, no. 3 (2018): 973–989.

Appiah, Anthony. *As If: Idealization and Ideals*. Cambridge, MA: Harvard University Press, 2017.

Aronson, Louise. "Examining Empathy." *The Lancet* 384, no. 9937 (2014): 16–17.

Bains, Sunny. "Nude Image Creates Feelings of Exclusion." *Electronic Engineering Times* (May 26, 1997): 45.

Banet-Weiser, Sarah. *Empowered: Popular Feminism and Popular Misogyny*. Durham, NC: Duke University Press, 2018.

Barad, Karen. *Meeting the Universe Halfway: Quantum Physics and the Entanglement of Matter and Meaning*. Durham, NC: Duke University Press, 2007.

Barnes, Colin. "Understanding the Social Model of Disability." In *Routledge Handbook of Disability Studies*, edited by Nick Watson, Alan Roulstone, and Carol Thomas, 12–29. London: Routledge, 2012.

Barrows, Howard S. "An Overview of the Uses of Standardized Patients for Teaching and Evaluating Clinical Skills." *Academic Medicine* 68 no. 6 (1993): 443–451.

Barrows, Howard S. "Simulated Patients in Medical Training." *Canadian Medical Association Journal* 98 (1968): 674–676.

Barrows, Howard S. *Simulated Patients (Programmed Patients): The Development and Use of a New Technique in Medical Education*. Springfield, IL: Charles C. Thomas, 1971.

Barrows, Howard S. *Simulated Patient Training: Acute Paralysis of Both Legs in a Young Woman* (video). Chapel Hill, NC: Health Sciences Consortium, 1988.

Barrows, Howard S. "Simulation in Medical Education." *Caduceus* 13, no. 2 (1997): 2–4.

Barrows, Howard S., and Stephen Abrahamson. "The Programmed Patient: A Technique for Appraising Student Performance in Clinical Neurology." *Journal of Medical Education* 39 (1964): 802–805.

Barrows, Howard S., Paul R. Patek, and Stephen Abrahamson. "Introduction of the Living Human Body in Freshman Gross Anatomy." *Medical Education* 2, no. 1 (1968): 33–35.

Barry, Andrew. "Materialist Politics: Metallurgy." In *Political Matter: Technoscience, Democracy, and Public Life*, edited by Bruce Braun and Sarah Whatmore, 89–110. Minneapolis: University of Minnesota Press, 2010.

Baxandall, Michael. *Painting and Experience in Fifteenth-Century Italy: A Primer in the Social History of Pictorial Style*. Oxford: Oxford University Press, 1988.

Bekiempis, Victoria. "MIT Scientist Resigns over Emails Discussing Academic Linked to Epstein." *The Guardian*, September 17, 2019. https://www.theguardian.com/education/2019/sep/17/mit-scientist-emails-epstein.

Benjamin, Ruha. *Race after Technology: Abolitionist Tools for the New Jim Code*. Cambridge, UK: Polity, 2019.

Berger, John. *Ways of Seeing*. London: Penguin Books, 1972.

Berlant, Lauren. "The Commons: Infrastructures for Troubling Times." *Environment and Planning D: Society and Space* 34, no. 3 (2016): 393–419.

Berlant, Lauren. *The Queen of America Goes to Washington City: Essays on Sex and Citizenship*. Durham, NC: Duke University Press, 1997.

Bivens, Rena. "The Gender Binary Will Not Be Deprogrammed: Ten Years of Coding Gender on Facebook." *New Media & Society* 9, no. 6 (2017): 880–898.

Bix, Amy Sue. *Girls Coming to Tech!* Cambridge, MA: MIT Press, 2014.

Bose, Dipan, Maria Segui-Gomez, and Jeff R. Crandall. "Vulnerability of Female Drivers Involved in Motor Vehicle Crashes: An Analysis of US Population at Risk." *American Journal of Public Health* 101 (December 2011): 2368–2373.

Boskin, Michael J., Ellen L. Dulberger, Robert J. Gordon, Zvi Griliches, and Dale W. Jorgenson. "Consumer Prices, the Consumer Price Index, and the Cost of Living." *Journal of Economic Perspectives* 12, no. 1 (1998): 3–26.

Bowden. Mark. *Black Hawk Down: A Story of Modern War*. New York: Signet, 1999.

Bowker, Geof. "How to Be Universal: Some Cybernetic Strategies, 1943–70." *Social Studies of Science* 23, no. 1 (1993): 107–127.

Bowker, Geoffrey C. *Memory Practices in the Sciences*. Cambridge, MA: MIT Press, 2005.

Bowker, Geoffrey C. *Science on the Run: Information Management and Industrial Geophysics at Schlumberger, 1920–1940*. Cambridge, MA: MIT Press, 1994.

Bowker, Geoffrey C., Karen Baker, Florence Millerand, and David Ribes. "Toward Information Infrastructure Studies: Ways of Knowing in a Networked Environment." In *International Handbook of Internet Research*, edited by Jeremy Hunsinger, Lisbeth Klastrup, and Matthew M. Allen, 97–117. Dordrecht, Netherlands: Springer, 2010.

Bowker, Geoffrey C., and Susan Leigh Star. *Sorting Things Out: Classification and Its Consequences*. Cambridge, MA: MIT Press, 1999.

Broomberg, Adam, and Oliver Chanarin. *To Photograph the Details of a Dark Horse in Low Light*. Gallery TPW R&D and Scotiabank Contact Photography Festival, 2013. http://gallerytpw.ca/rd/broomberg-chanarin/billboard-locations/.

Broussard, Meredith. *Artificial Unintelligence: How Computers Misunderstand the World*. Cambridge, MA: MIT Press, 2018.

Brown, Janelle. "Playmate Meets Geeks Who Made Her a Net Star." *Wired*, May 1997. http://archive.wired.com/culture/lifestyle/news/1997/05/4000.

Browne, Simone. *Dark Matters*. Durham, NC: Duke University Press, 2015.

Buolamwini, Joy, and Timnit Gebru. "Gender Shades: Intersectional Accuracy Disparities in Commercial Gender Classification." *Conference on Fairness, Accountability, and Transparency*, *PMLR* 81 (2018): 77–91.

Bureau International des Poids et Mesures. *SI Brochure: The International System of Units (SI)*. 8th ed. Paris: Stedi Media, 2006.

Burke, Kenneth. *Permanence and Change: An Anatomy of Purpose*. Los Altos, CA: Hermes Publications, 1954.

Burrington, Ingrid. *Networks of New York: An Illustrated Field Guide to Urban Internet Infrastructure*. Brooklyn: Melville House Printing, 2016.

Busch, Lawrence. *Standards: Recipes for Reality*. Cambridge, MA: MIT Press, 2011.

Butler, Judith. *Bodies That Matter: On the Discursive Limits of "Sex."* New York: Routledge, 1993.

Butler, Judith. "Performative Acts and Gender Constitution: An Essay in Phenomenology and Feminist Theory." *Theatre Journal* 40, no. 4 (1988): 519–531.

Callon, Michel. "Society in the Making: The Study of Technology as a Tool for Sociological Analysis." In *The Social Construction of Technological Systems: New Directions in the Sociology and History of Technology*, edited by Wiebe E. Bijker, Thomas P. Hughes, and Trevor Pinch, 83–103. Cambridge, MA: MIT Press, 1987.

Canguilhem, Georges. *The Normal and the Pathological*. Translated by Carolyn R. Fawcett. Brooklyn: Zone Books, 1989.

Cavell, Stanley. *In Quest of the Ordinary: Lines of Skepticism and Romanticism*. Chicago: Chicago University Press, 1994.

Chang, Hasok. *Inventing Temperature: Measurement and Scientific Progress*. New York: Oxford University Press, 2004.

Chemaly, Soraya. *Rage Becomes Her*. New York: Atria, 2018.

Chun, Wendy Hui Kyong. *Control and Freedom: Power and Paranoia in the Age of Fiber Optics*. Cambridge, MA: MIT Press, 2006.

Chun, Wendy Hui Kyong. *Programmed Visions: Software and Memory*. Cambridge, MA: MIT Press, 2011.

Chun, Wendy Hui Kyong. "Unbearable Witness: Toward a Politics of Listening." *differences: A Journal of Feminist Cultural Studies* 11, no. 1 (1999): 112–149.

Clough, Patricia T. "The Affective Turn: Political Economy, Biomedia and Bodies." *Theory, Culture & Society* 25, no. 1 (2008): 1–22.

Cockburn, Cynthia. "The Material of Male Power." *Feminist Review* 9, no. 1 (1981): 41–58.

Coleman, Gabriella. *Coding Freedom: The Ethics and Aesthetics of Hacking*. Princeton, NJ: Princeton University Press, 2012.

Combs, C. Donald. "Humans as Models." In *Modeling and Simulation in the Medical and Health Sciences*, edited by John Sokolowski and Catherine Banks, 85–108. Hoboken, NJ: John Wiley & Sons, 2011.

Comité International des Poids et Mesures (CIPM), Procès-Verbaux des Séances de 1882. Paris: Gauthier-Villars, 1883.

Comptes Rendus des Séances de la Première Conférence Générale des Poids et Mesures, Réunie a Paris en 1889. Paris: Guathier-Villars et Fils, Imprimeures-Libraires, 1890.

Corbman, Rachel. "The Scholars and the Feminists: The Barnard Sex Conference and the History of the Institutionalization of Feminism." *Feminist Formations* 27, no. 3 (2015): 49–80.

Cornfeld, Li. "Babes in Tech Land: Expo Labor as Capitalist Technology's Erotic Body." *Feminist Media Studies* 18, no. 2 (2018): 205–220.

Costanza-Chock, Sasha. *Design Justice: Community-Led Practices to Build the Worlds We Need*. Cambridge, MA: MIT Press, 2020.

Cottrell, Janet. "I'm a Stranger Here Myself: A Consideration of Women in Computing." *ACM SIGUCCS User Services Conference* 20 (1992): 71–76.

Cowan, T. L. "Digital Hygiene: A Metaphor of Dirty Proportions." Accessed March 4, 2019. http://www.drecollab.org/digital-hygiene-a-metaphor-of-dirty-proportions/.

Crease, Robert P. *World in the Balance: The Historic Quest for an Absolute System of Measurement*. New York: W. W. Norton, 2012.

Criado-Perez, Caroline. *Invisible Women: Exposing Data Bias in a World Designed for Men*. New York: Harry N. Abrams, 2019.

Crosland, Maurice. "The Congress on Definitive Metric Standards, 1798–1799: The First International Scientific Conference?" *Isis* 60, no. 2 (1969): 226–231.

Cullen Dunn, Elizabeth. "Standards without Infrastructure." In *Standards and Their Stories: How Quantifying, Classifying, and Formalizing Practices Shape Everyday Life*, edited by Martha Lampland and Susan Leigh Star, 118–121. Ithaca, NY: Cornell University Press, 2009.

Cumpson, Peter, and Sano Naoko. "Stability of Reference Masses V: UV/Ozone Treatment of Gold and Platinum Surfaces." *Metrologia* 50, no. 1 (2013): 27–36.

Cumpson, Peter, and M. P. Seah. "Stability of Reference Masses I: Evidence for Possible Variations in the Mass of Reference Kilograms Arising from Mercury Contamination." *Metrologia* 31, no. 1 (1994): 21–26.

Daniels, Jessie. "Cloaked Websites: Propaganda, Cyber-racism and Epistemology in the Digital Era." *New Media & Society* 11, no. 5 (2009): 659–683.

Daston, Lorraine, ed. *Things That Talk: Object Lessons from Art and Science*. Brooklyn: Zone Books, 2004.

Daston, Lorraine, and Peter Galison. *Objectivity*. Brooklyn: Zone Books, 2007.

Davidson, Stuart. "A Review of Surface Contamination and the Stability of Standard Masses." *Metrologia* 40, no. 6 (2003): 324–338.

Davis, Angelique M., and Rose Ernst. "Racial Gaslighting." *Politics, Groups, and Identities* 7, no. 4 (2019): 761–774.

Davis, Lennard J. *Enforcing Normalcy: Disability, Deafness, and the Body*. London: Verso, 1995.

Davis, Richard. "The SI Unit of Mass." *Metrologia* 40, no. 6 (2003): 299–305.

Davis, R. S. "Recalibration of the U.S. National Prototype Kilogram." *Journal of Research of the National Bureau of Standards* 90, no. 4 (1985): 263–283.

DeNardis, Laura. *Opening Standards: The Global Politics of Interoperability*. Cambridge, MA: MIT Press, 2011.

DeNardis, Laura. *Protocol Politics: The Globalization of Internet Governance*. Cambridge, MA: MIT Press, 2009.

DeNardis, Laura. "The Social Stakes of Interoperability." *Science* 337, no. 6101 (2012): 1454–1455.

de Podesta, Michael. "The Measure of Science: Redefining the Kilogram." Presentation, the Royal Institution, London, October 22, 2018.

Derrida, Jacques. "Declarations of Independence." *New Political Science* 7, no. 1 (1986): 7–15.

Dewey, John. *The Essential Dewey*, edited by Larry A. Hickman and Thomas M. Alexander. Bloomington: Indiana University Press, 1998.

Doane, Mary Ann. "Screening the Avant-Garde Face." In *The Question of Gender*, edited by Judith Butler and Elizabeth Weed, 206–229. Bloomington: Indiana University Press, 2011.

Dosekun, Simidele. *Fashioning Postfeminism: Spectacular Femininity and Transnational Culture*. Urbana: University of Illinois Press, 2020.

Douglas, Mary. *How Institutions Think*. Syracuse, NY: Syracuse University Press, 1986.

Douglas, Mary. *Purity and Danger: An Analysis of Concepts of Pollution and Taboo*. New York: Routledge, 2002.

Dourish, Paul. *The Stuff of Bits: An Essay on the Materialities of Information*. Cambridge, MA: MIT Press, 2017.

Downey, Greg. "Making Media Work: Time, Space, Identity, and Labor in the Analysis of Information and Communication Infrastructures." In *Media Technologies: Essays on Communication, Materiality, and Society*, edited by Tarleton Gillespie, Pablo Boczkowski, and Kirsten Foot, 141–165. Cambridge, MA: MIT Press, 2014.

Downey, Greg. "Virtual Webs, Physical Technologies, and Hidden Workers: The Spaces of Labor in Information Internetworks." *Technology and Culture* 42, no. 2 (2001): 209–235.

Dreyfuss, Henry. *The Measure of Man: Human Factors in Design*. 2nd ed. New York: Whitney Library of Design, 1967.

Duarte, Marisa Elena. *Network Sovereignty: Building the Internet across Indian Country*. Seattle: University of Washington Press, 2017.

Duchamp, Marcel. *The Writings of Marcel Duchamp*. New York: Da Capo Press, 1989.

Du Gay, Paul, Stuart Hall, Linda Janes, Anders Koed Madsen, Hugh Mackay, and Keith Negus. *Doing Cultural Studies: The Story of the Sony Walkman*. London: SAGE, 2013.

Dunbar-Hester, Christina. *Hacking Diversity: The Politics of Inclusion in Open Technology Cultures*. Princeton, NJ: Princeton University Press, 2020.

Durkheim, Émile. *The Elementary Forms of Religious Life*. Translated by Karen E. Fields. New York: Free Press, 1995.

Dutton, Michael, Sanjay Seth, and Leela Gandhi. "Plumbing the Depths: Toilets, Transparency and Modernity." *Postcolonial Studies* 5, no. 2 (2002): 137–142.

Dyer, Richard. *White: Essays on Race and Culture*. London: Routledge, 1997.

Dyer, Richard. "White." *Screen* 29 (Fall 1988): 44–64.

Edwards, Paul N. "Infrastructure and Modernity: Force, Time, and Social Organization in the History of Sociotechnical Systems." In *The History of Sociotechnical Systems: Modernity and Technology*, edited by Thomas Misa, Philip Brey, and Andrew Feenberg, 185–226. Cambridge, MA: MIT Press.

Edwards, Paul N. *A Vast Machine: Computer Models, Climate Data, and the Politics of Global Warming*. Cambridge, MA: MIT Press, 2010.

Eismann, Michael T. "Farewell, Lena." *Optical Engineering* 57, no. 12 (2018): 120101.

Enloe, Cynthia. *Maneuvers: The International Politics of Militarizing Women's Lives*. Berkeley: University of California Press, 2000.

Ensmenger, Nathan. "'Beards, Sandals, and Other Signs of Rugged Individualism': Masculine Culture within the Computing Professions." *Osiris* 30, no. 1 (2015): 38–65.

Ensmenger, Nathan. *The Computer Boys Take Over: Computers, Programmers, and the Politics of Technical Expertise*. Cambridge, MA: MIT Press, 2010.

Ensmenger, Nathan. "Is Chess the Drosophila of Artificial Intelligence? A Social History of an Algorithm." *Social Studies of Science* 42, no. 1 (2012): 5–30.

Eubanks, Virginia. *Automating Inequality*. New York: St. Martin's Press, 2018.

Feiner, John R., John W. Severinghaus, and Philip E. Bickler, "Dark Skin Decreases the Accuracy of Pulse Oximeters at Low Oxygen Saturation: The Effects of Oximeter Probe Type and Gender." *Anesthesia & Analgesia* 105, no. 6 (2007): S18–S23.

Fink, Donald G., and NTSC. *Color Television Standards; Selected Papers and Records*. New York: McGraw-Hill, 1955.

Finn, Megan. *Documenting Aftermath: Information Infrastructures in the Wake of Disasters*. Cambridge, MA: MIT Press, 2018.

Fleming, Sandford. *Time-Reckoning for the Twentieth Century*. Washington, DC: Smithsonian, 1889.

Flyverbom, Mikkel. *The Digital Prism*. Cambridge: Cambridge University Press, 2019.

Ford, Lisa. *Settler Sovereignty: Jurisdiction and Indigenous People in America and Australia, 1786–1836*. Cambridge, MA: Harvard University Press, 2010.

Ford, Matt. "Use It or Lose It?" *Atlantic Monthly*, May 30, 2017.

Forrester, John. "If p Then What? Thinking in Cases." *History of the Human Sciences* 9, no. 3 (1996): 1–25.

Foucault, Michel. *The Birth of the Clinic*. Translated by A. M. Sheridan. London: Routledge, 2003[1973].

Foucault, Michel. *The Order of Things: An Archaeology of the Human Sciences*. Translated by A. M. Sheridan. London: Routledge, 2002[1970].

Francis, George William, *The Dictionary of the Arts, Sciences, and Manufactures* (London: W. Brittain, 1846).

Fraser, Nancy. *Justice Interruptus: Critical Reflections on the "Postsocialist" Condition*. New York: Routledge, 1997.

Freedman, David, Robert Pisani, and Roger Purves. *Statistics*. New York: W. W. Norton, 2007.

Frenkel, Karen A. "Women and Computing." *Communications of the ACM* 33, no. 11 (1990): 34–46.

Frosh, Paul. *The Image Factory: Consumer Culture, Photography and the Visual Content Industry*. Oxford, UK: Berg Publishers, 2003.

Fuller, Matthew. *Media Ecologies: Materialist Energies in Art and Technoculture*. Cambridge, MA: MIT Press, 2005.

Gaboury, Jacob. *Image Objects: An Archaeology of 3D Computer Graphics, 1965–1979*. PhD dissertation, New York University, 2015.

Gaboury, Jacob. "A Queer History of Computing." *Rhizome*, 2013. https://rhizome.org/editorial/2013/feb/19/queer-computing-1/.

Galison, Peter. *Einstein's Clocks and Poincaré's Maps: Empires of Time*. New York: W. W. Norton, 2003.

Galison, Peter. "Image of Self." In *Things That Talk*, edited by Lorraine Daston, 257–294. Brooklyn: Zone Books.

Galloway, Alexander R. *Protocol: How Control Exists after Decentralization*. Cambridge, MA: MIT Press, 2004.

Gano, Selam Jie. "Remove Richard Stallman." *Medium*. Accessed September 12, 2019. https://medium.com/@selamjie/remove-richard-stallman-fec6ec210794.

Ganz Blythe, Sarah, and Edward D. Powers. *Looking at Dada*. New York: Museum of Modern Art, 2006.

Garcia, Chris. "Robots Are a Few of My Favorite Things." *Computer History Museum*. June 17, 2015. https://computerhistory.org/blog/robots-are-a-few-of-my-favorite-things-by-chris-garcia/?key=robots-are-a-few-of-my-favorite-things-by-chris-garcia.

Gardner, John. "The Many Faces of the Reasonable Person." *Law Quarterly Review* 131 (2015): 563–584.

Garland Thomson, Rosemary. *Extraordinary Bodies: Figuring Physical Disability in American Culture and Literature*. New York: Columbia University Press, 1997.

Gerrard, Ysabel, and Helen Thornham. "Content Moderation: Social Media's Sexist Assemblages." *New Media & Society* 22, no. 7 (2020): 1266–1286.

Gewerts, Ken. "A Bevy of Unknown Beauties," *Harvard Gazette*, July 21, 2005, http://news.harvard.edu/gazette/2005/07.21/00-girls.html.

Gieryn, Thomas. "What Buildings Do." *Theory and Society* 31, no.1 (2002): 35–74.

Gillespie, Tarleton. *Custodians of the Internet: Platforms, Content Moderation and the Hidden Decisions That Shape Social Media*. New Haven, CT: Yale University Press, 2018.

Gillespie, Tarleton. *Wired Shut: Copyright and the Shape of Digital Culture*. Cambridge, MA: MIT Press, 2007.

Girard, G. "The Washing and Cleaning of Kilogram Prototypes at the BIPM." *BIPM Internal Report,* 1990.

Gitelman, Lisa. *Always Already New*. Cambridge, MA: MIT Press, 2006.

Gitelman, Lisa, and Virginia Jackson. "Introduction." In *"Raw Data" Is an Oxymoron*, edited by Lisa Gitelman, 1–14. Cambridge, MA: MIT Press, 2013.

Glenn, Russell W., Jody Jacobs, Brian Nichiporuk, Christopher Paul, Barbara Raymond, Randall Steeb, and Harry J. Thie. *Preparing for the Proven Inevitable: An Urban Operations Training Strategy for America's Joint Force*. Santa Monica, CA: RAND Corporation, 2006.

Goffman, Erving. *Stigma: Notes on the Management of Spoiled Identity*. Englewood Cliffs, NJ: Prentice Hall, 1962.

Goodwin, Charles. "Professional Vision." *American Anthropologist* 96, no. 3 (1994): 606–633.

Gordon, Lewis. "Is the Human a Teleological Suspension of Man?" In *After Man, Towards the Human: Critical Essays on Sylvia Wynter*, edited by Anthony Bogues, 237–257. Kingston, Jamaica: Ian Randle, 2006.

Grosz, Barbara J., ed. *Report on Women in the Sciences at Harvard*. Cambridge, MA: Harvard University, 1991.

Guins, Raiford. *Edited Clean Version: Technology and the Culture of Control*. Minneapolis: University of Minnesota Press, 2008.

Hacking, Ian. *The Taming of Chance*. Cambridge: Cambridge University Press, 1990.

Hall, Rachel. *The Transparent Traveler*. Durham, NC: Duke University Press, 2015.

Hall, Stuart. "New Ethnicities." In *Stuart Hall: Critical Dialogues in Cultural Studies*, edited by Kuan-Hsing Chen and David Morley, 441–449. London: Routledge, 1996.

Hall, Stuart. "The Whites of Their Eyes: Racist Ideologies and the Media." In *Gender, Race, and Class in Media*, edited by Gail Dines and Jean M. Humez, 18–22. Thousand Oaks, CA: SAGE, 1995.

Hamraie, Aimi. *Building Access: Universal Design and the Politics of Disability*. Minneapolis: University of Minnesota Press, 2017.

Hasinoff, Amy Adele. *Sexting Panic: Rethinking Criminalization, Privacy, and Consent*. Urbana: University of Illinois Press, 2015.

Hecht, Gabrielle. "Interscalar Vehicles for an African Anthropocene: On Waste, Temporality, and Violence." *Cultural Anthropology* 33, no. 1 (2018): 109–141.

Hemmings, Clare. *Why Stories Matter: The Political Grammar of Feminist Theory*. Durham, NC: Duke University Press, 2011.

Henderson, Kathryn. "The Visual Culture of Engineers," *Sociological Review* 42, no. 1 (1994): 196–218.

Herrmann-Werner, Anne, Teresa Loda, Lisa M. Wiesner, Rebecca Sarah Erschens, Florian Junne, and Stephan Zipfel. "Is an Obesity Simulation Suit in an Undergraduate Medical Communication Class a Valuable Teaching Tool? A Cross-sectional Proof of Concept Study." *BJM Open* 9 (2019): e029738.

Hetherington, Kregg. "Surveying the Future Perfect: Anthropology, Development and the Promise of Infrastructure." In *Infrastructures and Social Complexity: A Companion*, edited by Penelope Harvey, Casper Bruun Jensen, and Atsuro Morita, 58–68. London: Routledge, 2016.

Hicks, Mar. *Programmed Inequality: How Britain Discarded Women Technologists and Lost Its Edge in Computing*. Cambridge, MA: MIT Press, 2017.

Hochschild, Arlie Russell. *The Managed Heart: Commercialization of Human Feeling*. 2nd ed. Berkeley: University of California Press, 2012.

Hodges, Brian David, and Nancy McNaughton. "Who Should Be an OSCE Examiner?" *Academic Psychiatry* 33, no. 4 (2009): 282–284.

Hoffman, Anna Lauren. "Terms of Inclusion: Data, Discourse, Violence." *New Media & Society* (September 2020). https://doi.org/10.1177/1461444820958725.

Hoffman, Kelly M., Sophie Trawalter, Jordan R. Axt, and M. Norman Oliver. "Racial Bias in Pain Assessment and Treatment Recommendations, and False Beliefs about Biological Differences between Blacks and Whites." *Proceedings of the National Academy of Sciences* 113, no. 16 (2016): 4296–4301.

Hogan, Mél. "Facebook Data Storage Centers as the Archive's Underbelly." *Television & New Media* 16, no. 1 (2015): 3–18.

Holt, Jennifer, and Patrick Vonderau. "Where the Internet Lives." In *Signal Traffic*, edited by Lisa Parks and Nicole Starosielski, 71–93. Urbana: University of Illinois Press, 2015.

hooks, bell. "Eating the Other: Desire and Resistance." In *Media and Cultural Studies: Keyworks*, edited by Meenakshi Gigi Durham and Douglas Kellner, 308–317. Malden, MA: Wiley, 2012.

Howley, Lisa D., Gayle Gliva-McConvey, Judy Thornton, and Association of Standardized Patient Educators. "Standardized Patient Practices: Initial Report on the Survey of US and Canadian Medical Schools." *Medical Education Online* 14 (2009): 4513. doi: 10.3885/meo.2009.F0000208.

Hsu, Hsuan L., and Martha Lincoln, "Biopower, *Bodies . . . the Exhibition,* and the Spectacle of Public Health." *Discourse* 29, no. 1 (2007): 15–34.

Hughes, Thomas P. "The Evolution of Large Technological Systems." In *The Social Construction of Technological Systems: New Directions in the Sociology and History of Technology*, edited by Wiebe E. Bijker, Thomas P. Hughes, and Trevor Pinch, 51–82. Cambridge, MA: MIT Press, 1987.

Hughes, Thomas P. *Networks of Power: Electrification in Western Society, 1880–1930*. Baltimore: Johns Hopkins University Press, 1983.

Hutchinson, Jamie. "Culture, Communication, and an Information Age Madonna." *IEEE Professional Communication Society Newsletter* 45, no. 3 (2001): 1, 5–7.

Igo, Sarah E. *The Averaged American: Surveys, Citizens, and the Making of a Mass Public*. Cambridge, MA: Harvard University Press, 2008.

Irigaray, Luce. *Speculum of the Other Woman*. Ithaca, NY: Cornell University Press, 1985.

Jackson, Steven. "Rethinking Repair." In *Media Technologies: Essays on Communication, Materiality, and Society*, edited by Tarleton Gillespie, Pablo Boczkowski, and Kirsten Foot, 221–239. Cambridge, MA: MIT Press, 2014.

Jackson, Steven, and David Ribes. "Data Bit Man: The Work of Sustaining a Long-Term Study." In *"Raw Data" Is an Oxymoron*, edited by Lisa Gitelman, 147–166. Cambridge, MA: MIT Press, 2013.

Jacyna, L. Stephen, and Stephen T. Casper. *The Neurological Patient in History*. Rochester, NY: University of Rochester Press, 2012.

Jain, Lochlann. *Injury: The Politics of Product Design and Safety Law in the United States*. Princeton, NJ: Princeton University Press, 2006.

Jamison, Leslie. *The Empathy Exams*. Minneapolis: Graywolf Press, 2014.

Jevons, William Stanley. *Principles of Science: A Treatise on Logic and Scientific Method*. New York: MacMillan, 1874.

Joint Committee for Guides in Metrology (JCGM). *International Vocabulary of Metrology*. 3rd ed. (2008). https://www.bipm.org/utils/common/documents/jcgm/JCGM_200_2012.pdf.

Kafer, Alison. *Feminist, Queer, Crip*. Bloomington; Indiana University Press, 2013.

Kafka, Franz. *The Complete Stories*, edited by Nahum N. Glatzer. New York: Schocken, 1971.

Kahn, Ellison. "A Trimestrial Potpourri." *South African Law Journal* 102, no. 1 (1985): 184–190.

Kahn, Robert E. "Oral History Interview with Robert E. Kahn." Minneapolis: Charles Babbage Institute, University of Minnesota Digital Conservancy, 1989. http://purl.umn.edu/107380.

Katz, Elihu, and Paul F. Lazarsfeld, *Personal Influence: The Part Played by People in the Flow of Mass Communication*. Piscataway, NJ: Transaction Publishers, 1955.

Kern, Stephen. *The Culture of Time and Space 1880–1918*. Cambridge, MA: Harvard University Press, 1983.

Kinstler, Linda. "Finding Lena, the Patron Saint of JPEGs." *Wired*, January 31, 2019. https://www.wired.com/story/finding-lena-the-patron-saint-of-jpegs/.

Kirschenbaum, Matthew. *Mechanisms: New Media and the Forensic Imagination*. Cambridge, MA: MIT Press, 2008.

Kittler, Friedrich. *Gramophone, Film, Typewriter*. Palo Alto, CA: Stanford University Press, 1999.

Kittler, Friedrich. *Optical Media*. Cambridge, UK: Polity, 2010.

Kohler, Robert. *Lords of the Fly: Drosophila Genetics and the Experimental Life*. Chicago: University of Chicago Press, 1994.

Kosofsky, Eve Sedgwick. *Between Men: English Literature and Male Homosocial Desire*. New York: Columbia University Press, 1985.

Krajewski, Markus. *The Server*. Translated by Ilinca Iurascu. New Haven, CT: Yale University Press, 2019.

Kuhn, Thomas. "Second Thoughts on Paradigms." In *The Essential Tension: Selected Studies in Scientific Tradition and Change*, edited by Frederick Suppe, 293–319. Chicago: University of Chicago Press, 1977.

Kuhn, Thomas. *The Structure of Scientific Revolutions*. 3rd ed. Chicago: University of Chicago Press, 1996.

Lampland, Martha, and Susan Leigh Star, eds. *Standards and Their Stories: How Quantifying, Classifying, and Formalizing Practices Shape Everyday Life*. Ithaca, NY: Cornell University Press, 2009.

Larkin, Brian. "The Politics and Poetics of Infrastructure." *Annual Review of Anthropology* 42, no. 1 (2013): 327–343.

Latour, Bruno. "Mixing Humans and Nonhumans Together: The Sociology of a Door-Closer." *Social Problems* 35, no. 3 (June 1988): 298–310.

Latour, Bruno. "On Technical Mediation." *Common Knowledge* 3, no. 2 (1994): 29–64.

Latour, Bruno, and Steve Woolgar. *Laboratory Life: The Construction of Scientific Facts*. Princeton, NJ: Princeton University Press, 1979.

Lehmann, Ann-Sophie. "Taking the Lid off the Utah Teapot towards a Material Analysis of Computer Graphics." *Zeitschrift für Medien-und Kulturforschung*, no. 1 (2012): 169–184.

Levine, Adam I., and Mark H. Swartz. "Standardized Patients: The 'Other' Simulation." *Journal of Critical Care* 23, no. 2 (2008): 179–184.

Leys, Ruth. *The Ascent of Affect*. Chicago: University of Chicago Press, 2017.

Lide, David R., ed. *A Century of Excellence in Measurements, Standards, and Technology: A Chronicle of Selected NBS/NIST Publications 1901–2000*. Washington, DC: U.S. Department of Congress, 2001.

Light, Jennifer S. "When Computers Were Women." *Technology and Culture* 40, no. 3 (1999): 455–483.

Long-Bellil, Linda, Kenneth L. Robey, Catherine L. Graham, Paula M. Minihan, Suzanne C. Smelzer, Paul Kahn, and Alliance for Disability in Health Care Education. "Teaching Medical Students about Disability: The Use of Standardized Patients." *Academic Medicine* 86, no. 9 (2011): 1163–1170.

Lowe, David G. "Object Recognition from Local Scale-Invariant Features." *Proceedings of the Seventh IEEE International Conference on Computer Vision*, vol. 2 (1999): 1150–1157.

Luotonen, Ari, and Kevin Altis. "World-Wide Web Proxies." *Computer Networks and ISDN Systems* 24, no. 2 (1994): 1–8.

Lynch, Robin. "Man Scans: The Matter of Expertise in Art and Technology Histories." *RACAR* (Spring 2021, forthcoming).

Lynd, Robert S., and Helen Merrell Lynd. *Middletown: A Study in American Culture*. New York: Harcourt, Brace, Jovanovich, 1956[1929].

Lynd, Robert S., and Helen Merrell Lynd. *Middletown in Transition: A Study in Cultural Conflicts*. Boston: Houghton Mifflin Harcourt, 1982[1937].

Mackenzie, Donald. *Inventing Accuracy*. Cambridge, MA: MIT Press, 1990.

Maguire, Ryan. "The Ghost in the MP3." 2014. http://ryanmaguiremusic.com/theghostinthemp3.html.

Makoul, Gregory. "Essential Elements of Communication in Medical Encounters: The Kalamazoo Consensus Statement." *Academic Medicine* 76, no. 4 (2001): 390–393.

Markham, James M. "Mob-Influenced Businesses Would Fill a List from A to Z, Officials Here Say." *New York Times*, August 19, 1972.

Marsh, Gerald. "Skirting Human Error: The Navy's Missile Launch System." *Bulletin of Atomic Scientists 43*, no. 1 (1987): 38–39.

Marvin, Carolyn. *When Old Technologies Were New: Thinking about Electric Communication in the Late Nineteenth Century*. New York: Oxford University Press, 1988.

Mattern, Shannon. "Maintenance and Care." *Places Journal*, November 2018. https://placesjournal.org/article/maintenance-and-care/.

McKelvey, Fenwick. *Internet Daemons: Digital Communications Possessed*. Minneapolis: University of Minnesota Press, 2018.

McKinney, Cait. "Crisis Infrastructures: AIDS Activism Meets Internet Regulation." In *AIDS and the Distribution of Crises*, edited by Jih-Fei Cheng, Alexandra Juhasz, and Nishant Shahani, 162–182. Durham, NC: Duke University Press, 2020.

McKinney, Cait. *Information Activism: A Queer History of Lesbian Media Technologies*. Durham, NC: Duke University Press, 2020.

McRuer. Robert. "Compulsory Able-Bodiedness and Queer/Disabled Existence." In *The Disability Studies Reader*, 3rd ed., edited by Lennard Davis. London: Routledge, 2010: 383–392.

McTiernan, John (dir.). *The Hunt for Red October*. Los Angeles: Paramount, 1990.

Melas, Natalie. *All the Difference in the World: Postcoloniality and the Ends of Comparison*. Palo Alto, CA: Stanford University Press, 2007.

Melnick, Donald E., Gerard F. Dillon, and David B. Swanson. "Medical Licensing Examinations in the United States." *Journal of Dental Education* 66, no. 5 (2002): 595–599.

Melzack, Ronald. "The McGill Pain Questionnaire: Major Properties and Scoring Methods." *Pain* 1, no. 3 (1975): 277–299.

Mendel, David. *Proper Doctoring: A Book for Patients and Their Doctors*. New York: New York Review Books Classics, 2013[1984].

Mills, Mara. "Do Signals Have Politics? Inscribing Abilities in Cochlear Implants." In *The Oxford Handbook of Sound Studies*, edited by Trevor Pinch and Karin Bijsterveld, 320–46. New York: Oxford University Press, 2011.

"Models Who Imitate Patients: Paradise for Medical Students." *San Francisco Chronicle*, September 28, 1965.

Molderings, Herbert. *Duchamp and the Aesthetics of Chance: Art as Experiment*. New York: Columbia University Press, 2010.

Montfort, Nick, and Ian Bogost. *Racing the Beam: The Atari Video Computer System*. Cambridge, MA: MIT Press, 2009.

Moran, Mayo. "The Reasonable Person: A Conceptual Biography in Comparative Perspective." *Lewis & Clark Law Review* 14 (2010): 1233–1283.

Mowlabocus, Sharif. *Gaydar Culture*. London: Ashgate/Routledge, 2010.

Muehlmann, Shaylih. "Clandestine Infrastructures: Illicit Connectivities in the US–Mexico Borderlands." In *Infrastructure, Environment, and Life in the Anthropocene*, edited by Kregg Hetherington, 45–65. Durham, NC: Duke University Press, 2019.

Mullaney, Thomas S. "The Moveable Typewriter: How Chinese Typists Developed Predictive Text during the Height of Maoism." *Technology and Culture* 53, no. 4 (2012): 777–814.

Mulvin, Dylan. "The Media of High-Resolution Time: Temporal Frequencies as Infrastructural Resources." *Information Society* 33, no. 5 (2017): 282–290.

Mulvin, Dylan. "Media Prophylaxis: Night Modes and the Politics of Preventing Harm." *Information & Culture* 53, no. 2 (2018): 175–202.

Mulvin, Dylan, and Jonathan Sterne. "Scenes from an Imaginary Country: Test Images and the American Color Television Standard." *Television & New Media* 17, no. 1 (2016): 21–43.

Munson, David C. "A Note on Lena." *IEEE Transactions on Image Processing* 5, no. 1 (1996): 3.

Murphy, Michelle. *The Economization of Life*. Durham, NC: Duke University Press, 2017.

Murphy, Michelle. *Sick Building Syndrome and the Problem of Uncertainty: Environmental Politics, Technoscience, and Women Workers*. Durham, NC: Duke University Press, 2006.

Murray, Susan. *Bright Signals*. Durham, NC: Duke University Press, 2018.

Myers West, Sarah. "Censored, Suspended, Shadowbanned: User Interpretations of Content Moderation on Social Media Platforms." *New Media & Society* 20, no. 11 (2018): 4366–4383.

Nakamura, Lisa. *Digitizing Race: Visual Cultures of the Internet*. Minneapolis: University of Minnesota Press, 2008.

Nakamura, Lisa. "Indigenous Circuits: Navajo Women and the Racialization of Early Electronic Manufacture." *American Quarterly* 66, no. 4 (2014): 919–941.

Nietzsche, Friedrich. *The Portable Nietzsche*. Translated by Walter Kaufmann. New York: Penguin Books, 1982.

Noble, David. *America by Design: Science, Technology, and the Rise of Corporate Capitalism*. New York: Knopf, 1982.

Noble, Safiya Umoja. *Algorithms of Oppression: How Search Engines Reinforce Racism*. New York: NYU Press, 2018.

Nooney, Laine. "A Pedestal, a Table, a Love Letter: Archaeologies of Gender in Videogame History." *Game Studies* 13, no. 2 (2013). http://gamestudies.org/1302/articles/nooney.

"A Note on the Lena Image." *Nature Nanotechnology* 13, no. 12 (2018): 1087.

Nowak, Peter. *Sex, Bombs, and Burgers: How War, Pornography, and Fast Food Have Shaped Modern Technology*. Guilford, CT: Rowman & Littlefield, 2011.

Nyamnjoh, Francis B. "Black Pain Matters: Down with Rhodes." *Pax Academica* 1, no. 2 (2015): 47–70.

Olesko, Kathryn M. "The Meaning of Precision: The Exact Sensibility in Early-Nineteenth-Century Germany." In *The Values of Precision*, edited by M. N. Wise, 103–134. Princeton, NJ: Princeton University Press, 1995.

O'Riordan, Kate, and David J. Phillips, eds. *Queer Online: Media Technology and Sexuality*. Bern, Switzerland: Peter Lang, 2007.

Paasonen, Susanna, Kylie Jarrett, and Ben Light. *NSFW: Sex, Humor, and Risk in Social Media*. Cambridge, MA: MIT Press, 2019.

Pachirat, Thomas. *Every Twelve Seconds*. New Haven, CT: Yale University Press, 2011.

Parks, Lisa, and Nicole Starosielski, eds. *Signal Traffic: Critical Studies of Media Infrastructures*. Urbana: University of Illinois Press, 2015.

Pasquale, Frank. *The Black Box Society: The Secret Algorithms That Control Money and Information*. Cambridge, MA: Harvard University Press, 2015.

Pearce, Christopher, and Steve Trumble. "Computers Can't Listen: Algorithmic Logic Meets Patient Centeredness." *Australian Family Physician* 35, no. 6 (2006): 439–442.

Peirce, Charles S. *Philosophical Writings of Peirce*, edited by Justus Buchler. New York: Dover Publications, 1955.

Peters, John Durham. "Witnessing." *Media, Culture & Society* 23, no. 6 (2001): 707–723.

Pinch, Trevor. "'Testing—One, Two, Three . . . Testing!': Toward a Sociology of Testing." *Science, Technology, & Human Values* 18, no. 1 (1993): 25–41.

Pinch, Trevor, and Wiebe E. Bijker. "The Social Construction of Facts and Artefacts: Or How the Sociology of Science and the Sociology of Technology Might Benefit Each Other." *Social Studies of Science* 14, no. 3 (1984): 399–441.

Planck, Max. "Über Irreversible Strahlungsvorgänge" [On Irreversible Radiation Processes]. *Annalen der Physik* 306, no. 1 (1900): 69–122.

Plantin, Jean-Christophe. "Data Cleaners for Pristine Datasets: Visibility and Invisibility of Data Processors in Social Science." *Science, Technology, & Human Values* 44, no. 1 (2019): 52–73.

Polanyi, Michael. *The Tacit Dimension*. New York: Doubleday, 1966.

Poovey, Mary, and Kevin R. Brine. "From Measuring Desire to Quantifying Expectations: A Late Nineteenth-Century Effort to Marry Economic Theory and Data." In *"Raw Data" Is an Oxymoron*, edited by Lisa Gitelman, 61–88. Cambridge, MA: MIT Press, 2013.

Porter, Roy. "The Patient's View: Doing Medical History from Below." *Theory and Society* 14, no. 2 (1985): 175–198.

Porter, Theodore. *The Rise of Statistical Thinking, 1820–1900*. Princeton, NJ: Princeton University Press, 1988.

Pow, Whitney. "Outside of the Folder, the Box, the Archive." *ROMchip* 1, no. 1 (2019). https://romchip.org/index.php/romchip-journal/article/view/76.

Powers, Kerns H. "Techniques for Increasing the Picture Quality of NTSC Transmissions in Direct Satellite Broadcasting." *IEEE Journal on Selected Areas in Communications* 3, no. 1 (1985): 57–64.

Pratt, William K. "A Bibliography on Television Bandwidth Reduction Studies." *IEEE Transactions on Information Theory* 13, no. 1 (1967): 114–115.

Pratt, William K. *Digital Image Processing: PIKS Scientific Inside.* Hoboken, NJ: Wiley-Interscience, 2007.

Pratt, William K. *Introduction to Digital Image Processing.* Boca Raton, FL: Taylor & Francis, 2014.

Pratt, William K. *USCEE Report #411: Semi-annual Technical Report Covering Research Activity during the Period 3 August 1971 to 29 February 1972.* Los Angeles: Signal and Image Processing Institute (SIPI), February 1972.

Pratt, William K. *USCIPI Report #660: Semi-annual Technical Report Covering Research Activity during the Period 1 September 1975 to 31 March 1976.* Los Angeles: Signal and Image Processing Institute (SIPI), March 1976.

Pratt, William K., and Harry C. Andrews. *Transform Processing and Coding of Images.* Los Angeles: Signal and Image Processing Institute (SIPI), 1969.

Prince, Stephen. *Classical Film Violence: Designing and Regulating Brutality in Hollywood Cinema, 1930–1968.* New Brunswick, NJ: Rutgers University Press, 2003.

Quetelet, Lambert Adolphe Jacques. *A Treatise on Man and the Development of His Faculties.* Edited by T. Smibert. Translated by R. Knox. Cambridge: Cambridge University Press, 2013.

Quinn, Terry. *From Artefacts to Atoms: The BIPM and the Search for Ultimate Measurement Standards.* Oxford: Oxford University Press, 2011.

Rader, Karen Ann. *Making Mice: Standardizing Animals for American Biomedical Research, 1900–1955.* Princeton, NJ: Princeton University Press, 2004.

Rankin, Joy Lisi. *A People's History of Computing in the United States.* Cambridge, MA: Harvard University Press, 2018.

Rankin, William J. "Infrastructure and The International Governance of Economic Development, 1950–1965." In *Internationalization of Infrastructures*, edited by Jean-François Auger, Jan Jaap Bouma, and Rolf Künneke, 61–75. Delft, Netherlands: Delft University of Technology, 2009.

Rault, Jas. "Tricks of Transparency in Colonial Modernity." Digital Research Ethics Collaboratory (DREC). http://www.drecollab.org/tricks-of-transparency/.

Rentschler, Carrie. "Witnessing: US Citizenship and the Vicarious Experience of Suffering." *Media, Culture & Society* 26, no. 2 (2004): 296–304.

Report of the MIT Committee on Family and Work. Cambridge, MA: MIT Press, 1990.

Rettberg, Jill Walker. "Machine Vision as Viewed through Art: Hostile Other or Part of Ourselves?" Paper presented at Post-Screen Festival: PSF2016, Lisbon, November 17–18, 2016.

Rheingold, Howard. *The Virtual Community: Homesteading on the Electronic Frontier*. Cambridge, MA: MIT Press, 2000.

Richard, Philippe, Hao Fang, and Richard Davis. "Foundation for the Redefinition of the Kilogram." *Metrologia* 53, no. 5 (2016): A6.

Riley, Donna, and Gina L. Sciarra. "'You're All a Bunch of Fucking Feminists': Addressing the Perceived Conflict between Gender and Professional Identities Using the Montreal Massacre." *Proceedings of the 36th ASEE/IEEE Frontiers in Education Conference* (2006): 19–24.

Roberts, Lawrence G. *Machine Perception of Three-Dimensional Solids*. New York: Garland Publishing, 1980.

Roberts, Lawrence G. "Picture Coding Using Pseudo-Random Noise." *IEEE Transactions on Information Theory* 8, no. 2 (1962): 145–154.

Roberts, Sarah T. *Behind the Screen: Content Moderation in the Shadows of Social Media*. New Haven, CT: Yale University Press, 2019.

Robertson, Craig. *The Filing Cabinet: A Vertical History of Information*. Minneapolis: University of Minnesota Press, 2021.

Robertson, Craig. "Learning to File: Reconfiguring Information and Information Work in the Early Twentieth Century." *Technology and Culture* 58, no. 4 (2017): 955–981.

Robertson, Craig. *The Passport in America: The History of a Document*. New York: Oxford University Press, 2010.

Rosenblatt, Frank. "The Perceptron: A Probabilistic Model for Information Storage and Organization in the Brain." *Psychological Review* 65, no. 6 (1958): 386–408.

Rosenfeld, Azriel. "From Image Analysis to Computer Vision: An Annotated Bibliography, 1955–1979." *Computer Vision and Image Understanding* 84, no. 2 (2001): 298–324.

Roth, Lorna. "Looking at Shirley, the Ultimate Norm: Colour Balance, Image Technologies, and Cognitive Equity." *Canadian Journal of Communication* 34, no. 1 (2009): 111–136.

Ruíz, Elena. "Cultural Gaslighting." *Hypatia* 35, no. 4 (2020): 687–713.

Russell, Andrew L. *Open Standards and the Digital Age: History, Ideology, and Networks*. Cambridge: Cambridge University Press, 2014.

Salmon, Nathan. "How to Measure the Standard Metre." *Proceedings of the Aristotelian Society* 88 (1987): 193–217.

Salomon, David. *Data Compression: The Complete Reference*. London: Springer, 2007.

Sanders, Heywood T. "What Infrastructure Crisis?" *Public Interest*, no. 110 (1993): 3–18.

Scarry, Elaine. *The Body in Pain: The Making and Unmaking of the World*. New York: Oxford University Press, 1985.

Schechner, Richard. *Between Theatre and Anthropology*. Philadelphia: University of Pennsylvania Press, 1985.

Schechner, Richard. *Performance Studies: An Introduction*. New York: Routledge, 2002.

Schneider, Rebecca. *Performing Remains: Art and War in Times of Theatrical Reenactment*. London: Routledge, 2011.

Schwartz Cowan, Ruth. *More Work for Mother*. New York: Basic Books, 1983.

Scott, James C. *Seeing Like a State: How Certain Schemes to Improve the Human Condition Have Failed*. New Haven, CT: Yale University Press, 1998.

Scott, Joan W. "Multiculturalism and the Politics of Identity." *October* 61 (1992): 12–19.

Scott, Ridley (dir.). *Black Hawk Down*. Los Angeles: Columbia Pictures, 2001.

Seaver, Nick. "Algorithms as Culture: Some Tactics for the Ethnography of Algorithmic Systems." *Big Data & Society* 4, no. 2 (2017): https://doi.org/10.1177/2053951717738104.

Sewell, Philip W. *Television in the Age of Radio: Modernity, Imagination, and the Making of a Medium*. New Brunswick, NJ: Rutgers University Press, 2014.

Shakespeare, Tom, and Nicholas Watson. "The Social Model of Disability: An Outdated Ideology?" *Research in Social Science and Disability* 2 (2002): 9–28.

Shapin, Steven. "Cordelia's Love: Credibility and the Social Studies of Science." *Perspectives on Science* 3, no. 3 (1995): 255–275.

Shcherbina, Anna, C. Mikael Mattsson, Daryl Waggott, Heidi Salisbury, Jeffrey W. Christie, Trevor Hastie, et al. "Accuracy in Wrist-Worn, Sensor-Based Measurements of Heart Rate and Energy Expenditure in a Diverse Cohort." *Journal of Personalized Medicine* 7, no. 2 (2017): 3–14.

Shell, Hanna Rose. *Hide and Seek: Camouflage, Photography, and the Media of Reconnaissance*. New York: Zone Books, 2012.

Siebers, Tobin. "Disability as Masquerade." *Literature and Medicine* 23, no. 1 (2004): 1–22.

Siegel, Greg. *Forensic Media*. Durham, NC: Duke University Press, 2014.

Simpson, Audra. "The Ruse of Consent and the Anatomy of 'Refusal': Cases from Indigenous North America and Australia." *Postcolonial Studies* 20, no.1 (2017): 18–33.

Slaton, Amy, and Janet Abbate. "The Hidden Lives of Standards: Technical Prescriptions and the Transformation of Work in America." In *Technologies of Power*, edited by Michael Thad Allen and Gabrielle Hecht, 95–144. Cambridge, MA: MIT Press, 2001.

Slayton, Rebecca. "Revolution and Resistance: Rethinking Power in Computing History." *IEEE Annals of the History of Computing* 30, no. 1 (2008): 96–97.

Spade, Dean. *Normal Life: Administrative Violence, Critical Trans Politics, and the Limits of Law*. Durham, NC: Duke University Press, 2011.

Spertus, Ellen. "Why Are There So Few Female Computer Scientists?" MIT Artificial Intelligence Laboratory, Cambridge, MA, 1991.

Spillers, Hortense J. "Mama's Baby, Papa's Maybe: An American Grammar Book." *Diacritics* 17, no. 2 (1987): 65–81.

Srinivasan, Amia. "The Aptness of Anger." *Journal of Political Philosophy,* 26, no. 2 (2018): 123–144.

Stanislavski, Constantin. *An Actor Prepares.* London: Routledge, 1989.

Stapleford, Thomas A. *The Cost of Living in America: A Political History of Economic Statistics, 1880–2000.* Cambridge: Cambridge University Press, 2009.

Star, Susan Leigh. "The Ethnography of Infrastructure." *American Behavioral Scientist* 43, no. 3 (1999): 377–391.

Star, Susan Leigh, and Karen Ruhleder. "Steps toward an Ecology of Infrastructure: Design and Access for Large Information Spaces." *Information Systems Research Information Systems Research* 7, no. 1 (1996): 111–134.

Starosielski, Nicole. *The Undersea Network.* Durham, NC: Duke University Press, 2015.

Steedman, Carolyn. "Something She Called a Fever: Michelet, Derrida, and Dust." *American Historical Review* 106, no. 4 (2001): 1159–1180.

Steenson, Molly Wright. *Architectural Intelligence: How Designers and Architects Created the Digital Landscape.* Cambridge, MA: MIT Press, 2017.

Sterne, Jonathan. "Ballad of the Dork-o-Phone: Towards a Crip Vocal Technoscience." *Journal of Interdisciplinary Voice Studies* 4, no. 2 (2019): 179–189.

Sterne, Jonathan. *MP3: The Meaning of a Format.* Durham, NC: Duke University Press, 2012.

Sterne, Jonathan, and Dylan Mulvin. "The Low Acuity for Blue: Perceptual Technics and American Color Television." *Journal of Visual Culture* 13, no. 2 (2014): 118–138.

Steyerl, Hito. *How Not to Be Seen: A Fucking Didactic Educational .MOV File* (video), Museum of Modern Art, New York, 2013. https://www.moma.org/collection/works/181784.

Strathern, Marilyn. *Reproducing the Future: Anthropology, Kinship, and the New Reproductive Technologies.* London: Routledge, 1992.

Suchman, Lucy. "Configuring the Other: Sensing War through Immersive Simulation." *Catalyst: Feminism, Theory, Technoscience* 2, no. 1 (2016): 1–36.

Sun, Lena H. "Demand Is High for Pretend Patients." *Washington Post,* October 14, 2011.

Sutherland, Ivan. "Oral History Interview with Ivan Sutherland." Minneapolis: Charles Babbage Institute, University of Minnesota Digital Conservancy, 1989. http://purl.umn.edu/107642.

Swartz, Lana. *New Money: How Payment Became Social Media.* New Haven, CT: Yale University Press, 2020.

Swauger, Shea. "Software That Monitors Students during Tests Perpetuates Inequality and Violates Their Privacy." *MIT Technology Review,* August 7, 2020. https://www.technologyreview.com/2020/08/07/1006132/software-algorithms-proctoring-online-tests-ai-ethics/.

Taylor, Janelle S. "The Moral Aesthetics of Simulated Suffering in Standardized Patient Performances." *Culture, Medicine and Psychiatry* 35, no. 2 (2011): 134–162.

Temple University School of Medicine. "Standardized Patient Program: Questions and Answers about Working as a Standardized Patient for Temple University School of Medicine" (2013). https://medicine.temple.edu/sites/medicine/files/files/FAQs.pdf.

Terry, Richard, Erik Hiester, and Gary D. James. "The Use of Standardized Patients to Evaluate Family Medicine Resident Decision Making." *Family Medicine* 39, no. 4 (2007): 261–265.

Thomas, William, and Dorothy Thomas. *The Child in America: Behavior Problems and Programs*. New York: Alfred A. Knopf, 1928.

Thompson, Brian J. "Editorial: Copyright Problems." *Optical Engineering* 31, no. 1 (1992): 5.

Towns, Armond R. "Toward a Black Media Philosophy." *Cultural Studies* 34 (2020): 851–873.

Tremain, Shelley. "On the Subject of Impairment." In *Disability/Postmodernity: Embodying Disability Theory*, edited by Mairian Corker and Tom Shakespeare, 32–47. New York: Bloomsbury, 2002.

Turnbull, David. "The Ad Hoc Collective Work of Building Gothic Cathedrals with Templates, String, and Geometry." *Science, Technology, & Human Values* 18, no. 3 (1993): 315–340.

Turner, Fred. *From Counterculture to Cyberculture: Steward Brand, the Whole Earth Network, and the Rise of Digital Utopianism*. Chicago: University of Chicago Press, 2006.

Tushnet, Rebecca. "Power without Responsibility: Intermediaries and the First Amendment." *George Washington Law Review* 76, no. 4 (2008): 986–1016.

Vaihinger, Hans. *The Philosophy of 'As If.'* Translated by C. K. Ogden. New York: Harcourt Brace, 1924.

Veblen, Thorstein. *The Instinct of Workmanship and the State of the Industrial Arts*. New York: Macmillan Company, 1914.

Vertesi, Janet. *Seeing Like a Rover: How Robots, Teams, and Images Craft Knowledge of Mars*. Chicago: University of Chicago Press, 2015.

Wallace, Peggy. "Following the Threads of an Innovation: The History of Standardized Patients in Medical Education." *Caduceus* 13, no. 2 (1997): 5–28.

Warner, Michael. *The Trouble with Normal: Sex, Politics, and the Ethics of Queer Life*. New York: Free Press, 1999.

Washington, Harriet A. *Medical Apartheid: The Dark History of Medical Experimentation on Black Americans from Colonial Times to the Present*. New York: Doubleday Books, 2006.

Wendell, Susan. *The Rejected Body: Feminist Philosophical Reflections on Disability*. New York: Routledge, 1996.

Wernimont, Jacqueline. *Numbered Lives: Life and Death in Quantum Media*. Cambridge, MA: MIT Press: 2019.

Williams, Christopher. *Kodak Three Point Reflection Guide © 1968 Eastman Kodak Company, 1968 (Meiko laughing), Vancouver, B.C., April 6, 2005*, Whitney Museum of American Art, New York. https://whitney.org/collection/works/27531.

Williams, Linda. *Hard Core: Power, Pleasure, and the "Frenzy of the Visible."* Berkeley: University of California Press, 1989.

Wilson, Benjamin, Judy Hoffman, and Jamie Morgenstern, "Predictive Inequity in Object Detection." *arXiv preprint:1902.11097* (2019).

Winner, Langdon. "Do Artifacts Have Politics?" In *The Whale and the Reactor*, 19–39. Chicago: University of Chicago Press, 1986.

Winslett, Marianne, ed. *Final Report of the Committee on the Status of Women Graduate Students and Faculty in the College of Engineering.* University of Illinois at Urbana-Champaign, Urbana, IL, June 30, 1993.

Winston, Brian. *Technologies of Seeing: Photography, Cinematography and Television.* London: British Film Institute, 1997.

Wise, M. Norton, ed. *The Values of Precision.* Princeton, NJ: Princeton University Press, 1997.

Wittgenstein, Ludwig. *Philosophical Investigations.* Oxford, UK: B. Blackwell, 1953.

Woodward, Bob, and Carl Bernstein. *All the President's Men.* New York: Simon & Schuster, 1974.

Woodward, Bob, and Carl Bernstein. "Funds Laundered." *Washington Post*, July 11, 1974.

Yates, JoAnne, and Craig N. Murphy. *Engineering Rules: Global Standard Setting since 1880.* Baltimore: Johns Hopkins University Press, 2019.

Yue, Genevieve. "The China Girl on the Margins of Film." *October* 153 (2015): 96–116.

Index

Note: Page numbers in *italics* refer to figures and tables.

Pixl test images, 90, *91*
Planck, Max, 50
Planck constant, 37, 211n8
Platform moderation, 128, 227n37
Playboy Enterprises, 117–120, 127, 129–130
Playboy Enterprises, Inc. v. Russ Hardenburgh,
    127–129
Playboy Enterprises, Inc. v. Webbworld Inc, 127
Playboy magazine
  issues at SIPI, 76–77, 80, 98, 102, 111,
    133
  Lena centerfold, 73
  photographs used as test images, 98–99.
    See also Lena test image
Podesta, Michael de, 35–38, 64
Politics of representation, 6, 19, 28. *See also*
    White skin in test images
Politics of standing-in, 138
Pornography
  in computer science environments, 132,
    135–136
  feminist debates, 226n23
  internet, 125–129
  mainstreaming of, 111
  new media adoption, 125. *See also* Images
    of nude women; Lena test image
Porousness of proxies, 27, 39, 146, 201
  and culture, 86
  IPK examples, 69–70
  Lena test image, 102, 143
  mediating function, 20
  standardized patients, 146
  test image examples, 78
Pratt, William, 98, 102–103, 131,
    223n79
Professional vision, 81, 219n20
Prototype objects. *See* Standard objects
Prototypical whiteness, 86–88, 96, 120
Proxies, 4–5
  accessibility-security paradox, 55
  averageness, 25–27
  bodily, 8

critique, 171
environmental interactions, 70
as fictions, 25, 27
following, 200–201
humans sustaining, 11–12
influence, 202
legal system, 14
living, 10–11
maintenance needs, 70–71, 143, 146, 198
materialized hypotheses, 47–48
mediation, 5, 20
ordinariness, 19
power, 30, 33
problem appearance, 33
reality creation, 28–29
reality practices, 22
requirements for functioning, 53
revealing norms, 9
selection consequences, 15–17
spectacularity, 183
taking for granted, 6–7, 19
ubiquity of, 7
unjust, 139
Proxification, 5, 14
Proxy servers, 12

Quetelet, Lambert "Adolphe" Jacques, 25–26
Quinn, Terry, 38

Racial gaslighting, 157
Racism
  denial of Black pain, 155–157
  infrastructure, 196–197
  medical systems, 156–157
  standardized patient programs, 157–158
  training data, 84–86
  voter suppression, 45. *See also* Test image
    racism; Whiteness; White skin in test
    images
Rault, Jas, 196
Raven, Lucy, 189, 237n15
Reasonable person proxies, 14–15

Stefan Höhne, *New York City Subway: The Invention of the Urban Passenger*

Timothy Moss, *Conduits of Berlin: Remaking the City through Infrastructure, 1920–2020*

Blake Atwood, *Underground: The Secret Life of Videocassettes in Iran*

Huub Dijstelbloem, *Borders as Infrastructure: The Technopolitics of Border Control*

Claude Rosental, *The Demonstration Society*

Dylan Mulvin, *Proxies: The Cultural Work of Standing In*